RAND COUNTERINSURGENCY STUDY • VOLUME 4

Counterinsurgency in Afghanistan

Seth G. Jones

D1531967

Prepared for the Office of the Secretary of Defense

Approved for public release; distribution unlimited

 NATIONAL DEFENSE RESEARCH INSTITUTE

The research described in this report was prepared for the Office of the Secretary of Defense (OSD). The research was conducted in the RAND National Defense Research Institute, a federally funded research and development center sponsored by the OSD, the Joint Staff, the Unified Combatant Commands, the Department of the Navy, the Marine Corps, the defense agencies, and the defense Intelligence Community under Contract W74V8H-06-C-0002.

Library of Congress Cataloging-in-Publication Data

Jones, Seth G., 1972–
    Counterinsurgency in Afghanistan / Seth G. Jones.
        p. cm. — (Rand counterinsurgency study ; v.4)
    Includes bibliographical references.
    ISBN 978-0-8330-4133-3 (pbk. : alk. paper)
      1. Afghanistan—Politics and government—2001– 2. Counterinsurgency—
Afghanistan. 3. Afghan War, 2001——Commando operations. 4. Counterinsurgency.
I. Title.

    DS371.4.J66 2008
    958.104'7—dc22

                                2008016686

Cover photo by Army Staff Sgt. Michael L. Casteel

The RAND Corporation is a nonprofit research organization providing objective analysis and effective solutions that address the challenges facing the public and private sectors around the world. RAND's publications do not necessarily reflect the opinions of its research clients and sponsors.

RAND® is a registered trademark.

Cover design by Stephen Bloodsworth

Published 2008 by the RAND Corporation
1776 Main Street, P.O. Box 2138, Santa Monica, CA 90407-2138
1200 South Hayes Street, Arlington, VA 22202-5050
4570 Fifth Avenue, Suite 600, Pittsburgh, PA 15213-2665
RAND URL: http://www.rand.org/
To order RAND documents or to obtain additional information, contact
Distribution Services: Telephone: (310) 451-7002;
Fax: (310) 451-6915; Email: order@rand.org

Preface

This book examines counterinsurgency operations in Afghanistan following the overthrow of the Taliban regime in 2001. It is based on repeated trips to Afghanistan, Pakistan, and India in 2004, 2005, 2006, 2007, and 2008. It focuses on the early stages of the insurgency—from 2002 until 2008—and examines why and how the insurgency began. It also draws lessons from the broader literature on counterinsurgency warfare and provides recommendations to help the United States develop capabilities and improve performance in future counterinsurgency operations. The focus of this research is on the U.S. military. However, since the actions of other U.S. government agencies, countries, international organizations such as the United Nations (UN), and nongovernmental organizations (NGOs) are obviously important, they are discussed where appropriate. The results should be of interest to a broad audience of policymakers and academics concerned with counterinsurgency and the related issues of state-building, nation-building, and stability operations.

This research was sponsored by the Office of the Secretary of Defense and conducted within the International Security and Defense Policy Center of the RAND National Defense Research Institute, a federally funded research and development center sponsored by the Office of the Secretary of Defense, the Joint Staff, the Unified Combatant Commands, the Department of the Navy, the Marine Corps, the defense agencies, and the defense Intelligence Community. For more information on RAND's International Security and Defense Policy Center, contact the Director, James Dobbins. He can be reached by email at James_Dobbins@rand.org; by phone at 703-413-1100, extension 5134; or by mail at the RAND Corporation, 1200 South Hayes Street, Arlington, VA 22202-5050. More information about RAND is available at www.rand.org.

Contents

Figures

Table

Summary

Following the initial success of U.S. and Afghan forces in overthrowing the Taliban regime in 2001, an increasingly violent insurgency began to develop. A mixed group of insurgents comprised of the Taliban, Hezb-i-Islami, the Haqqani network, foreign fighters, local tribes, and criminal organizations began a sustained effort to overthrow the Afghan government. U.S. and coalition efforts in Afghanistan offer a useful opportunity to assess what works—and what does not—in counterinsurgency warfare. This study examines the beginning of the insurgency in Afghanistan and asks three major questions. First, what was the nature of the insurgency? Second, what factors have contributed to the rise of insurgencies more broadly and to the insurgency in Afghanistan in particular? Third, what capabilities should the U.S. military consider developing to improve its ability to wage effective counterinsurgency operations?

The core argument of this study is that the United States should focus its resources on developing capabilities that help improve the capacity of the *indigenous government* and its security forces to wage counterinsurgency warfare. It has not always done this well. The analysis of 90 insurgencies since 1945 in Chapter Two indicates that three variables are correlated with the success (and failure) of counterinsurgency efforts:

- capability of indigenous security forces, especially police
- local governance
- external support for insurgents, including sanctuary.

The U.S. military—along with other U.S. and coalition partners—is more likely to be successful in counterinsurgency warfare the more capable and legitimate the indigenous security forces are (especially the police), the better the capacity of the local government is, and the less external support to insurgents there is. The indigenous government and its forces have a greater chance of gaining, in Max Weber's words, a "monopoly of the legitimate use of physical force within a given territory."[1] In many cases, a significant direct intervention by U.S. military forces may undermine popular support and legitimacy. The United States is also unlikely to remain for the duration of most insurgencies: This study's assessment of 90 insurgencies indicates that it takes an average of 14 years to defeat insurgents once an insurgency develops.

In the Afghan insurgency, the competence—and, in some areas, incompetence—of the indigenous government and its security forces have been critical factors. This analysis suggests that success in Afghanistan hinges on three factors.

First is the ability of the United States and other international actors to help build competent and legitimate Afghan security forces, especially police, which was not accomplished during the early stages of the counterinsurgency. Repeated trips to the regional police training centers in Afghanistan, as well as interviews with police in the field, indicated that the Afghan National Police were corrupt, incompetent, underresourced, and often loyal to local commanders rather than to the central government. Indeed, the Afghan police received little attention and were a low priority in the early stages of the counterinsurgency. This was a mistake. The police are the primary arm of the government in a counterinsurgency because of their presence in local villages and districts. The U.S. military made significant changes in the police training program beginning in 2005 and 2006, but persistence is the key to police reform. Based on the low quality of Afghan police when the Taliban was overthrown in 2001, police reform in Afghanistan will take at least a decade.

---

[1]   Max Weber, "Politics as a Vocation," in H. H. Gerth and C. Wright Mills, eds., *From Max Weber: Essays in Sociology* (New York: Oxford University Press, 1958), p. 78.

Second, the United States and other international actors need to improve the quality of local governance, especially in rural areas of Afghanistan. Field research in the east and south showed that development and reconstruction did not reach most rural areas because of the deteriorating security environment. Even the Provincial Reconstruction Teams, which were specifically designed to assist in development and reconstruction projects, operated in pockets in the east and south because of security concerns. NGOs and state agencies, such as the U.S. Agency for International Development and the Canadian International Development Agency, were also not involved in reconstruction and development in many areas of the south and east. The irony in this situation is that rural areas, which were most at risk from the Taliban and where unhappiness with the slow pace of change was greatest among the population, received little assistance. The counterinsurgency in Afghanistan will be won or lost in the local communities of rural Afghanistan, not in urban centers such as Kabul. This means the counterinsurgency must find ways to reach these communities despite security concerns.

Third, the United States and other international actors need to eliminate the insurgents' support base in Pakistan. The failure to do so will cripple long-term efforts to stabilize and rebuild Afghanistan. Every successful insurgency in Afghanistan since 1979 enjoyed a sanctuary in Pakistan and assistance from individuals within the Pakistan government, such as the Frontier Corps and the Inter-Services Intelligence Directorate (ISI).

The Taliban and other insurgent groups enjoyed a sanctuary in the Federally Administered Tribal Areas and Balochistan Province. The Taliban regularly shipped arms, ammunition, and supplies into Afghanistan from Pakistan. Many suicide bombers came from Afghan refugee camps located in Pakistan, and improvised explosive device components were often smuggled across the Afghanistan-Pakistan border and assembled at safe houses in such provinces as Kandahar. The Taliban used roads such as Highway 4 in Kandahar Province to transport fighters and supplies between Afghanistan and Pakistan. And the leadership structure of most insurgent groups (e.g., the Taliban, Hezb-i-Islami, the Haqqani network, and al Qaeda) was based in

Pakistan. There is some indication that individuals within the Pakistan government—for example, within the Frontier Corps and the ISI—were involved in assisting insurgent groups. Solving this problem will require a difficult political and diplomatic feat: convincing the government of Pakistan to undermine the sanctuary on its soil.

This effort became more challenging with the rise of an insurgency in Pakistan by a range of militant groups, members of which assassinated Pakistani opposition leader Benazir Bhutto and conducted brazen attacks against the Pakistan army, ISI, and officials from other government agencies. Militants from Pakistan's border areas were also linked to a range of international terrorist attacks and plots, such as the July 2005 attacks on London's mass transit system, the foiled 2006 plot against transatlantic commercial aircraft flights, foiled plots in 2007 in Germany and Denmark, and the 2008 arrests of terrorist suspects in Spain. These developments indicated that the insurgency in Afghanistan had spread to neighboring Pakistan and required a regional solution.

Most policymakers—including those in the United States—repeatedly ignore or underestimate the importance of locals in counterinsurgency operations. Counterinsurgency requires not only the capability of the United States to conduct unconventional war, but, most importantly, the ability to shape the capacity of the indigenous government and its security forces. U.S. military and civilian efforts should focus on leveraging indigenous capabilities and building capacity. In some areas, such as air strikes and air mobility, this may be difficult. The recommendations in Chapter Seven cover eight functional areas: police, border security, ground combat, air strike and air mobility, intelligence, command and control, information operations, and civil-military affairs. In some of these areas, such as civil affairs, the U.S. military should not be the lead agency and will need to coordinate closely with other states, international organizations, and NGOs. Indeed, the success of any counterinsurgency campaign over the long run ultimately requires a combination of military, political, economic, and other efforts.

Acknowledgments

This book would not have been possible without the help of numerous individuals. The most significant are Ben Riley and Richard Higgins from the Office of the Secretary of Defense, whose support and vision allowed this research to happen. Nora Bensahel, James Dobbins, Ali Jalali, and Barnett Rubin provided excellent and frank reviews of earlier drafts, which greatly improved the overall quality of the book. At RAND, Farhana Ali, Cheryl Benard, Keith Crane, David Frelinger, David Gompert, John Gordon, Martin Libicki, Ed O'Connell, Bruce Pirnie, William Rosenau, and Obaid Younossi provided valuable information on Afghanistan and counterinsurgency operations. Several others also imparted useful information and comments about Afghanistan, Pakistan, and counterinsurgency more broadly. They include Daniel Byman, Christine Fair, Bruce Hoffman, and Robert Perito. Hekmat Karzai and his Centre for Conflict and Peace Studies in Kabul provided a wonderful opportunity to share ideas. Nathan Chandler provided key research support and collected data for many of the charts and graphs.

I owe a special debt of gratitude to those government officials from Afghanistan, Pakistan, India, the United States, Canada, Australia, and Europe who provided critical information about insurgents and counterinsurgency efforts and took time out of their busy schedules. Most did not want to be identified.

Key Afghan officials to whom I talked over the course of my research included Foreign Minister Rangin Dadfar Spanta, National Security Advisor Zalmai Rassoul, Minister of Interior Ali Jalali,

Ambassador Said Tayeb Jawad, General Ghulam Ghaws Naseri, National Security Council staff member Daoud Yaqub, and Deputy Minister of Justice Mohammad Qasim Hashimzai. Key U.S. officials with whom I spoke included Ambassador Ronald Neumann, LTG Karl Eikenberry, LTG David Barno, Ambassador Zalmay Khalilzad, MAJ GEN Craig P. Weston, LTC William R. Balkovetz, Jack Bell, Tom Berner, COL Paul Calbos, COL Joseph D. Celeski, Doug Climan, LTC David Duffy, Ray Fitzgerald, COL Walter Herd, Martin Hoffman, Andrew Mann, COL Gary Medvigy, Thomas A. Pastor, COL John Reardon, Marin Strmecki, Edward M. Staff, Ambassador William Taylor, Ken Thomas, Doug Wankel, and COL Mike Winstead. I am also grateful for the assistance of officials from Germany, Italy, the United Kingdom, and the United Nations who agreed to discuss counterinsurgency and Afghanistan with me. A special thanks to Christopher Alexander, Carlo Batori, Walter Dederichs, Ambassador Helmut Frick, Paul George, Ursula Müller, Larry Sampler, Ron Sandee, Alexandre Schmidt, Ambassador David Sproule, Ambassador Arif Lalani, and Ambassador Rainald Steck for their insights.

Abbreviations

ANA	Afghan National Army
ANP	Afghan National Police
CIA	Central Intelligence Agency
COIN	counterinsurgency
HF	high frequency
HUMINT	human intelligence
IED	improvised explosive device
ISI	Inter-Services Intelligence Directorate
NATO	North Atlantic Treaty Organization
NDS	National Directorate for Security (Afghanistan)
NGO	nongovernmental organization
PRT	Provincial Reconstruction Team
SIGINT	signals intelligence
SOFLAM	Special Operations Forces Laser Acquisition Marker
UN	United Nations
UNAMA	United Nations Assistance Mission in Afghanistan
USAID	U.S. Agency for International Development

CHAPTER ONE

Introduction

In 2001, the United States orchestrated a rapid military victory in Afghanistan. A combination of U.S. Special Operations and Central Intelligence Agency (CIA) forces, air power, and Afghan indigenous troops overthrew the Taliban regime in less than three months; U.S. forces suffered only a dozen casualties.[1] Some individuals involved in the operation argued that it revitalized the American way of war.[2] However, this initial success was quickly succeeded by the emergence of a prolonged insurgency as the Taliban, Hezb-i-Islami, the Haqqani network, foreign fighters, local militias, and criminal organizations began a sustained effort to overthrow the new Afghan government. This study defines an insurgency as a political-military campaign by nonstate actors seeking to overthrow a government or secede from a country through the use of unconventional—and sometimes conventional—military strategies and tactics.[3]

---

[1] On the overthrow of the Taliban regime, see Gary Schroen, *First In: An Insider's Account of How the CIA Spearheaded the War on Terror in Afghanistan* (New York: Ballantine Books, 2005); Stephen Biddle, *Afghanistan and the Future of Warfare: Implications for Army and Defense Policy* (Carlisle, Pa.: Strategic Studies Institute, U.S. Army War College, November 2002); Gary Berntsen and Ralph Pezzullo, *Jawbreaker: The Attack on Bin Laden and Al Qaeda* (New York: Crown Publishers, 2005); Bob Woodward, *Bush at War* (New York: Simon and Schuster, 2002).

[2] Henry A. Crumpton, "Intelligence and War: Afghanistan 2001–2002," in *Transforming U.S. Intelligence*, ed. Jennifer E. Sims and Burton Gerber (Washington, D.C.: Georgetown University Press, 2005), p. 177.

[3] On the definition of insurgency, see Central Intelligence Agency, *Guide to the Analysis of Insurgency* (Washington, D.C.: Central Intelligence Agency, n.d.), p. 2; *Department of*

This study asked three major questions. First, what was the nature of the insurgency in Afghanistan? Second, what factors have contributed to the rise of insurgencies more broadly and to the insurgency in Afghanistan in particular? Third, what capabilities should the U.S. Department of Defense consider developing to improve its ability to wage effective counterinsurgency operations? This chapter outlines the research effort, and then provides a brief outline of the book.

Research Design

The research design adopted is straightforward. It included conducting an exhaustive set of primary source interviews in Afghanistan, Pakistan, India, the United States, and Europe (including during multiple visits to Afghanistan in 2004, 2005, 2006, 2007, and 2008). These included conversations with several hundred government officials from the United States, Afghanistan, Pakistan, India, and the North Atlantic Treaty Organization (NATO), as well as staff from the United Nations (UN) and several nongovernmental organizations (NGOs). Finally, the research included a review and analysis of hundreds of government documents from the United States, Afghanistan, and coalition countries such as Germany and the United Kingdom, as well as transcripts and videos from the Taliban, Hezb-i-Islami, and al Qaeda. To supplement the research on Afghanistan, several researchers at RAND (including the author) built a data set of all of the 90 insurgencies that occured since 1945. Our goal was to identify the variables that can be correlated with the success and failure of insurgencies.

This research design offers a useful means for assessing U.S. counterinsurgency warfare capabilities because it provides an opportunity to examine what worked, what did not, and why.[4] But there are draw-

---

Defense Dictionary of Military and Associated Terms, Joint Publication 1-02 (Washington, DC: U.S. Department of Defense, 2001), p. 266.

[4] In particular, see Alexander L. George, "Case Studies and Theory Development: The Method of Structured, Focused Comparison," in Paul Gordon Lauren, ed., *Diplomacy: New Approaches in History, Theory, and Policy* (New York: Free Press, 1979), pp. 43–68.

backs to relying solely on a single case study such as this one.[5] A single case is a limited laboratory for identifying those capabilities that are effective across a range of counterinsurgencies, since conditions can vary across countries. For example, several factors beyond the quality of the indigenous government and its forces can impact the outcome of counterinsurgency: geography (such as mountainous terrain); degree of urbanization; ethnic, tribal, or religious fissures within the state; and economic conditions.[6] In a single case study, there is rarely variation across these factors. Afghanistan, for instance, has rugged mountains in much of the country; a large rural population; a range of ethnic and tribal groups; and poor economic conditions. Consequently, a study of Afghanistan cannot provide a definitive assessment of what types of military and nonmilitary capabilities might be useful against insurgents operating among a homogenous population living in the jungle or in major cities.[7] What may work in Afghanistan may not work in all other countries.

---

[5] On the costs and benefits of comparative case studies, see David Collier, "The Comparative Method: Two Decades of Change," in *Comparative Political Dynamics: Global Research Perspectives*, ed. Dankwart A. Rustow and Kenneth Paul Erickson (New York: Harper Collins, 1991), pp. 7–31; Charles C. Ragin, "Comparative Sociology and the Comparative Method," *International Journal of Comparative Sociology*, Vol. 22, Nos. 1–2 (March–June 1981), pp. 102–120; Charles Tilly, "Means and Ends of Comparison in Macrosociology," in Lars Mjoset and Frederik Engelstad, eds., *Comparative Social Research,* Vol. 16: *Methodological Issues in Comparative Social Science* (Greenwich, Conn.: JAI Press, 1997), pp. 43–53; Theda Skocpol and Margaret Somers, "The Uses of Comparative History in Macrosocial Inquiry," *Comparative Studies in Society and History*, Vol. 22, No. 2 (1980), pp. 174–197; Stephen Van Evera, *Guide to Methods for Students of Political Science* (Ithaca, N.Y.: Cornell University Press, 1997), pp. 49–88.

[6] James D. Fearon and David D. Laitin, "Ethnicity, Insurgency, and Civil War," *American Political Science Review*, Vol. 97, No. 1 (February 2003), pp. 75–90.

[7] See, for example, Gary King, Robert Keohane, and Sidney Verba, *Designing Social Inquiry: Scientific Inference in Qualitative Research* (Princeton, N.J.: Princeton University Press, 1994), pp. 208–230; John H. Goldthorpe, "Current Issues in Comparative Macrosociology: A Debate on Methodological Issues," in Mjoset and Engelstad, *Comparative Social Research,* Vol. 16, pp. 1–26; David Collier and James Mahoney, "Insights and Pitfalls: Selection Bias in Qualitative Research," *World Politics*, Vol. 49, No. 1 (October 1996), pp. 56–91.

Despite these factors, however, there are several reasons why a case study of Afghanistan is useful. First, the outcome of the insurgency in Afghanistan is of such intrinsic importance to the United States that its lessons are particularly important. The attacks in Washington, D.C., New York, and Pennsylvania on September 11, 2001, were planned in Afghanistan, and many of the hijackers received training there. Consequently, U.S. performance during the counterinsurgency campaign has significant implications for U.S. national security. As the 9/11 Commission Report concluded, a U.S. failure to stabilize Afghanistan would decrease U.S. security by allowing the country to become a safe haven for terrorists and criminals.[8] Second, a single case provides a good opportunity to carefully examine what worked and what did not, sometimes referred to as "process tracing."[9] It allows us to infer and test explanations of how U.S. capabilities and strategies affected counterinsurgency efforts—and why. As Alexander George and Timothy McKeown argue, case studies are useful in uncovering

> what stimuli the actors attend to; the decision process that makes use of these stimuli to arrive at decisions; the actual behavior that then occurs; the effect of various institutional arrangements on attention, processing, and behavior; and the effect of other variables of interest on attention, processing, and behavior.[10]

The focus of this research is on the U.S. military and its capabilities for conducting counterinsurgency warfare. The actions of the White House, the Department of State, the U.S. Agency for International Development (USAID), the CIA, and other U.S. government organizations are obviously critical during counterinsurgency opera-

---

[8]   *The 9/11 Commission Report: Final Report of the National Commission on Terrorist Attacks Upon the United States* (New York: W. W. Norton, 2004), pp. 369–371.

[9]   Alexander L. George and Timothy J. McKeown, "Case Studies and Theories of Organizational Decision Making," in *Advances in Information Processing in Organizations: A Research Annual*, Vol. 2, ed. Robert F. Coulam and Richard A. Smith (Greenwich, Conn.: JAI Press, 1985), pp. 34–41.

[10]   King, Keohane, and Verba, *Designing Social Inquiry*, pp. 226–228; George and McKeown, "Case Studies and Theories of Organizational Decision Making," p. 35.

tions. So are the actions of other states, international organizations, and NGOs. As David Galula argues, counterinsurgency operations "are essentially of a political nature." This means that "political action remains foremost throughout the war" and "every military move has to be weighed with regard to its political effects, and vice versa."[11] Nevertheless, the military plays a particularly critical role in counterinsurgency warfare—and will continue to do so in the future. While the focus of this research is on the role of the U.S. military and the development of its counterinsurgency capabilities, the role of other agencies is noted where appropriate. After all, the success of any counterinsurgency campaign over the long term requires a combination of political, economic, and military resources brought to bear by a variety of governmental and nongovernmental actors.

Outline

Chapter Two critiques some of the current arguments about counterinsurgency warfare and offers an alternative framework for understanding it. Chapter Three provides a brief overview of Afghanistan's "age of Insurgency" beginning in 1979. Chapter Four examines lessons that can be learned from the insurgents, including the Taliban, Hezb-i-Islami, foreign fighters, local tribes, and criminal organizations. Chapter Five outlines lessons from the Afghan government and its security forces, and Chapter Six examines lessons from the United States and coalition forces. The focus in these chapters is primarily on the strategic and operational level, rather than the tactical level. Chapter Seven pulls together lessons from the three sets of actors—insurgents, the Afghan government, and the U.S. military—and then outlines key capabilities for counterinsurgency warfare.

---

[11] David Galula, *Counterinsurgency Warfare: Theory and Practice* (St. Petersburg, Fla.: Hailer Publishing, 2005), p. 9.

CHAPTER TWO

Success in Counterinsurgency Warfare

Most military figures and policymakers—including those in the United States—underestimate the importance of the indigenous government and its security forces in counterinsurgency warfare. This chapter argues that the focus of the U.S. military should be *to improve the competence and legitimacy of indigenous actors to conduct counterinsurgency operations*. Achieving this goal involves increasing the capacity of indigenous security forces to wage military and nonmilitary operations, improving governance, and undermining external support for insurgents. These steps are critical in winning popular support and ensuring legitimacy for the indigenous government. This chapter begins by critiquing the current thinking on counterinsurgency warfare and then offers a sounder strategy for success.

The Fallacy of External Actors

One of the key challenges in waging effective counterinsurgency operations is understanding the variables that impact their success (or failure). Most assessments of counterinsurgency operations tend to ignore or downplay the role of indigenous forces and mistakenly focus on how to improve the capabilities of *outside forces* to directly defeat insurgents. This might include revising the U.S. military's organizational structure or increasing external resources (such as troops) to directly counter insurgents. This approach assumes the recipe for a successful counterinsurgency is adapting the U.S. military's capabilities so it can win the support of the local population and defeat insurgents. The problem with this approach is that it ignores or underestimates the most critical actor in a counterinsurgency campaign: the indigenous government and its security forces.

This mistake is common in the counterinsurgency literature. John Nagl argues, for example, that success in counterinsurgency operations is largely a function of an external military's ability to adapt its organizational structure and strategy to win the support of the local population and directly defeat insurgents. But he largely ignores the role of the indigenous government and its security forces.[1] In assessing the British performance in Malaya and the U.S. performance in Vietnam, Nagl concludes

> [T]he better performance of the British army in learning and implementing a successful counterinsurgency doctrine in Malaya (as compared to the American army's failure to learn and implement successful counterinsurgency doctrine in Vietnam) *is best explained by the differing organizational cultures of the two armies*; in short, that the British army was a learning institution and the American army was not.[2]

General Frank Kitson, who participated in several counterinsurgency campaigns in Africa, Europe, and Asia, similarly argues that a successful campaign needs to take into account three groups: the insurgent group's political structure, the insurgent group's military structure, and the population. Kitson argues that external forces need to focus on defeating the insurgent's political and military infrastructure and winning the support of the population.[3] However, he largely ignores the role of indigenous actors. In his study of French counterinsurgency operations, Roger Trinquier makes a similar mistake. He argues that the key to success is adapting the external military's ability to directly

---

[1] John A. Nagl, *Learning to Eat Soup with a Knife: Counterinsurgency Lessons from Malaya and Vietnam* (Chicago: University of Chicago Press, 2005), p. xiv. Nagl later conceded that his book pays little attention to working with—and through—indigenous forces.

[2] Nagl, *Learning to Eat Soup with a Knife*, p. xxii. Emphasis added.

[3] Frank Kitson, *Low Intensity Operations: Subversion, Insurgency and Peacekeeping* (London: Faber and Faber, 1971), p. 49.

defeat insurgent groups. The failure to adapt, he notes, was the main reason French forces were defeated in Indochina and Algeria: "The result of this shortcoming is that the army is not prepared to confront an adversary employing arms and methods the army itself ignores. It has, therefore, no chance of winning."[4] In addition, U.S. Army Colonel Timothy Deady argues that the United States was successful in the Philippines because of direct U.S. action. Using Mao Tse-tung's aphorism that insurgents are like fish that need a sea in which to swim, he concludes that

> American strategy effectively targeted both the insurgents' strategic and operational centers of gravity . . . . As American garrisons drained the local lakes, the insurgent fish became easier to isolate and catch.[5]

All of these works commit a similar fallacy: They ignore or downplay the role of the indigenous government and its security forces. This focus on winning counterinsurgency campaigns by improving the capabilities of external actors has become conventional wisdom among numerous military officials and counterinsurgency experts. However, such a strategy is misplaced. While improving the U.S. military's ability to directly counter insurgents may be necessary to a successful counterinsurgency campaign, it is not sufficient. In particular, it underestimates the importance of indigenous forces: Most counterinsurgency campaigns are not won or lost by external forces, but by indigenous forces. The quality of indigenous forces and government has significantly impacted the outcome of past counterinsurgencies.[6] Shaping a

---

[4] Roger Trinquier, *Modern Warfare: A French View of Counterinsurgency,* trans. Daniel Lee (New York: Praeger, 1964), p. 3.

[5] Timothy K. Deady, "Lessons from a Successful Counterinsurgency: The Philippines, 1899–1902," *Parameters*, Vol. XXXV, No. 1 (Spring 2005), p. 58.

[6] Daniel L. Byman, "Friends Like These: Counterinsurgency and the War on Terrorism," *International Security*, Vol. 31, No. 2 (Fall 2006), pp. 79–115; Daniel L. Byman, *Going to War with the Allies You Have: Allies, Counterinsurgency, and the War on Terrorism* (Carlisle, Pa.: U.S. Army War College, November 2005); James Corum, *Training Indigenous Forces in Counterinsurgency: A Tale of Two Insurgencies* (Carlisle, Pa.: U.S. Army War College, 2006).

successful counterinsurgency is not just a matter of adapting the organizational structure of an external military to unconventional war. It
requires an understanding of the nature of the local conflict and the
ability to shape the capacity of indigenous actors to conduct an effective counterinsurgency campaign. This includes a range of steps such as
effectively training police and improving governance capacity.

Indeed, there are dangers in focusing too heavily on a lead U.S.
role and improving U.S. military capabilities to directly act against
insurgents. First, U.S. forces are unlikely to remain for the duration
of any counterinsurgency effort, at least as a major combatant force.[7]
Insurgencies are usually of short duration only if the indigenous government collapses at an early stage. An analysis of all insurgencies since
1945 shows that successful counterinsurgency campaigns last for an
average of 14 years, and unsuccessful ones last for an average of 11
years. Many also end in a draw, with neither side winning. Insurgencies can also have long tails: Approximately 25 percent of insurgencies
won by the government and 11 percent won by insurgents last more
than 20 years.[8] Since indigenous forces eventually have to win the war
on their own, they must develop the capacity to do so. If they do not
develop this capacity, indigenous forces are likely to lose the war once
international assistance ends.[9] Second, indigenous forces usually know
the population and terrain better than external actors and are better
able to gather intelligence. Third, a lead U.S. role may be interpreted
by the population as an occupation, eliciting nationalist reactions that

---

[7]   Kimberly Marten Zisk, *Enforcing the Peace: Learning from the Imperial Past* (New York:
Columbia University Press, 2004); Amitai Etzioni, "A Self-Restrained Approach to Nation-
Building by Foreign Powers," *International Affairs*, Vol. 80, No. 1 (2004); Amitai Etzioni,
From Empire to Community: A New Approach to International Relations (New York: Palgrave
Macmillan, 2004); Stephen T. Hosmer, *The Army's Role in Counterinsurgency and Insurgency*
(Santa Monica, Calif: RAND Corporation, R-3947-A, 1990), pp. 30–31.

[8]   Unpublished RAND research for the U.S. Department of Defense. On time, also see
Galula, *Counterinsurgency Warfare*, p. 10.

[9]   On rentier states, see Barnett R. Rubin, *The Fragmentation of Afghanistan: State Formation
and Collapse in the International System* (New Haven, Conn.: Yale University Press, 2002),
pp. 81–105; Charles Tilly, *The Formation of National States in Western Europe* (Princeton,
N.J.: Princeton University Press, 1975); Hazem Beblawi and Giacomo Luciani, eds., *The
Rentier State* (New York: Croom Helm, 1987).

impede success.[10] Fourth, a lead indigenous role can provide a focus for national aspirations and show the population that they—and not foreign forces—control their destiny. Competent governments that can provide services to their population in a timely manner can best prevent and overcome insurgencies.

An Indigenous Lead

As Figure 2.1 highlights, insurgencies involve three sets of actors. The first are insurgents, which include those groups that adopt unconventional—and sometimes conventional—military strategies and tactics to overthrow an established national government or secede from it.[11] The second is the indigenous government, which includes the government's security forces, such as the army and police, as well as its governance capacity. Governance involves the provision of essential services to the population by a central authority in a timely manner, including health care, power, transportation infrastructure, and other basic services. The third group of actors comes from outside. These include states and nonstate entities, which can support the indigenous government or the insurgents. As explained in more detail below, outside actors can play a pivotal role in tipping the war in favor of insurgents

---

[10] David M. Edelstein, "Occupational Hazards: Why Military Occupations Succeed or Fail," *International Security*, Vol. 29, No. 1 (Summer 2004), p. 51.

[11] The *Department of Defense Dictionary of Military and Associated Terms* (Joint Publication 1-02) defines unconventional warfare as

> A broad spectrum of military and paramilitary operations, normally of long duration, predominantly conducted by indigenous or surrogate forces who are organized, trained, equipped, supported and directed in varying degrees by an external source. It includes guerrilla warfare, and other direct offensive, low visibility, covert, or clandestine operations, as well as the indirect activated of subversion, sabotage, intelligence activities and evasion and escape.

Consequently, conventional war refers to warfare conducted by using conventional military weapons—such as tanks and artillery—and battlefield tactics between two or more states in open confrontation. See, for example, *Department of Defense Dictionary of Military and Associated Terms*, p. 556.

Figure 2.1
A Counterinsurgency Framework

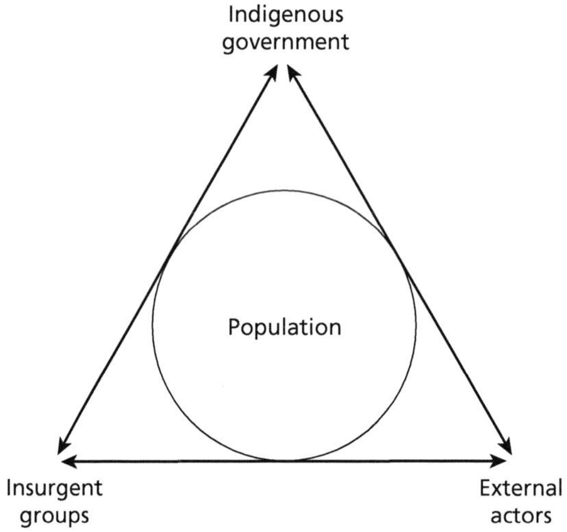

RAND MG595-2.1

or the government. However, outside actors alone can rarely win the war for either side.

Popular support is a common goal for all actors in an insurgency. Both winning support and preventing insurgents from gaining support are critical components of any counterinsurgency.[12] With popular support comes assistance—money, logistics, recruits, intelligence, and other aid—from the local population. Building on Mao Tse-tung's argument that the guerrilla must move among the people as a fish swims in the sea, British General Sir Frank Kitson argued that the population

---

[12] Bruce Hoffman, *Insurgency and Counterinsurgency in Iraq* (Santa Monica, Calif.: RAND Corporation, OP-127-IPC/CMEPP, 2004); U.S. Marine Corps, *Small Wars Manual* (Washington, D.C.: U.S. Government Printing Office, 1940); Julian Pagent, *Counter-Insurgency Campaigning* (London: Faber and Faber, 1967); Charles Simpson, *Inside the Green Berets: The First Thirty Years* (Novato, Calif.: Presidio Press, 1982); Robert J. Wilensky, *Military Medicine to Win Hearts and Minds: Aid to Civilians in the Vietnam War* (Lubbock, Tex.: Texas Tech University Press, 2004).

is a critical element in counterinsurgency operations as "this represents the water in which the fish swims."[13]

The Afghan insurgency can be understood using the framework in Figure 2.1. The Afghan government and such forces as the Afghan National Army (ANA) and Afghan National Police (ANP) are the primary indigenous counterinsurgency actors. The Taliban, Hezb-i-Islami, the Haqqani network, foreign fighters, criminal groups, and a host of Afghan and Pakistani tribal militias are the primary insurgent forces. There are two sets of external actors. The United States, NATO forces, and other international actors such as the United Nations are the primary external actors that support the Afghan government. The broad jihadist network, individuals within the Pakistan government, and Pakistani and Afghan tribes are the primary external actors that support the insurgents. In between the two sides are the Afghan and, to some degree, Pakistani populations, which lie at the center of insurgent and counterinsurgent efforts.

The population is particularly critical to insurgents because of their relative weakness. Insurgents generally cannot attack their opponents in a conventional manner, as the government forces are usually much more capable of waging conventional warfare. This asymmetry in power forces insurgents to carry the fight to an arena in which they have a better chance of success. To many insurgents, the population offers a level playing field. If insurgents manage to alienate the population from the government and acquire its active support, they are more likely to win the war. In the end, the exercise of political power depends on the tacit or explicit agreement of the population—or, at worst, on its submissiveness.[14]

Figure 2.2 illustrates the framework of the counterinsurgency campaign in Afghanistan. As the dotted lines indicate, outside actors such as the U.S. military are likely to play an indirect role over the long run by providing resources to the Afghan government. It is unlikely

---

[13] Kitson, *Low Intensity Operations*, p. 49. On counterinsurgency strategies, also see Colonel C. E. Callwell, *Small Wars: Their Principles and Practice*, 3rd ed. (Lincoln, Neb.: University of Nebraska Press, 1996), pp. 34–42; Galula, *Counterinsurgency Warfare*, pp. 17–42.

[14] Trinquier, *Modern Warfare*, p. 8; Galula, *Counterinsurgency Warfare:*, pp. 7–8.

Figure 2.2
A Framework for Afghanistan

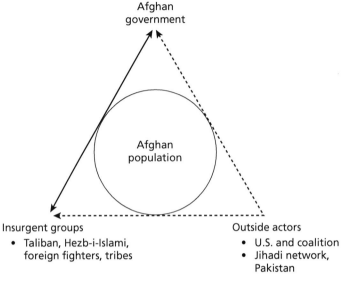

Afghan
government

Afghan
population

Insurgent groups
- Taliban, Hezb-i-Islami,
  foreign fighters, tribes

Outside actors
- U.S. and coalition
- Jihadi network,
  Pakistan

RAND *MG595-2.2*

that the insurgents will be defeated before the end of direct U.S. military intervention. As the solid line indicates, the long-term struggle is between the Afghan government and insurgent groups.

To identify the critical variables that have contributed to the success or failure of past insurgencies, RAND constructed a data set aggregating information on insurgencies occurring since World War II.[15] These insurgencies met the following three criteria: (1) they involved fighting between agents of (or claimants to) a state and nonstate groups who sought to take control of a government, take power in a region, or use violence to change government policies; (2) at least 1,000 individuals were killed over the course of the conflict, with a yearly average of at least 100; and (3) at least 100 were killed on both sides (including civilians attacked by rebels). These criteria resulted in a list of 90 insurgencies, which is provided in the appendix.

---

[15] Unpublished RAND research for the U.S. Department of Defense. On time, see also Galula, *Counterinsurgency Warfare*, p. 10.

Analysis of this database shows that several variables can be linked with the success (or failure) of counterinsurgency efforts:

- capability of indigenous security forces, especially police
- quality of local governance
- external support for insurgents, including sanctuary.

In addition, there are several other variables that correlate with the success of insurgencies, such as the type of terrain and the size of the population.[16] However, these factors are outside the control of external and indigenous actors, especially in the near term.

Security Forces

The capability of the government security forces to defeat insurgents and establish law and order is paramount to the success of any counterinsurgency. Insurgents are better able to survive and prosper if the security forces they oppose are relatively weak and lack legitimacy with the population. These forces may be badly financed and equipped, organizationally inept, corrupt, politically divided, and poorly informed about events at the local level.[17] Indigenous governments often rely on military and paramilitary forces to conduct a significant part of a counterinsurgency, since they may have more firepower than police to use against well-armed insurgent forces. There are numerous ideal characteristics of these forces at the tactical and operational levels, but several of the most important include a high level of initiative, good intelligence, integration across units and services, quality leadership, motivated soldiers, and the ability to learn and adapt during combat.[18] As Figure 2.3 illustrates, there is some correlation between government

---

[16] For example, mountainous terrain, larger populations, and lower per capita income levels increase the likelihood of insurgent success. But indigenous government and external powers such as the United States can do little about these variables. Galula, *Counterinsurgency Warfare*, pp. 37–38; Fearon and Laitin, "Ethnicity, Insurgency, and Civil War," pp. 83, 85.

[17] Fearon and Laitin, "Ethnicity, Insurgency, and Civil War."

[18] Byman, "Friends Like These," pp. 79–115; Byman, *Going to War with the Allies You Have*.

Figure 2.3
Competency of Security Forces and Success of Counterinsurgencies

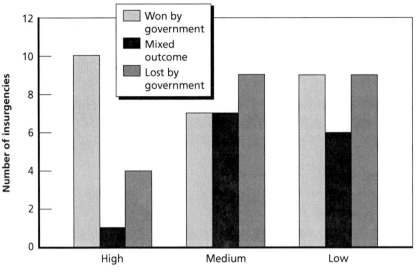

RAND *MG595-2.3*

competence at counterinsurgency and success.[19] Governments with competent security forces won in two-thirds of all completed insurgencies, but governments defeated less than a third of the insurgencies when their competence was medium or low.

While military and paramilitary forces play a key role, the police are perhaps the most critical component of indigenous forces. They are the primary arm of the government focused on internal security matters. Unlike the military, the police usually have a permanent presence in cities, towns, and villages; a better understanding of the threat environment in these areas; and better intelligence. This makes them

---

[19] We assessed the capability of government security forces by making a qualitative judgment about how competent their forces were in conducting counterinsurgency warfare. We tried to avoid the endogeneity problem of coding forces as competent if the government won—and incompetent if they lost. Rather, we relied on the judgments of area specialists and historians that covered each insurgency. Unpublished RAND research for the U.S. Department of Defense.

a direct target of insurgent forces, who often try to kill or infiltrate them. The mission of the police and other security forces should be to eliminate the insurgent organization—the command structure, guerrillas, logistics support, and financial and political support—from the midst of the population.[20] An effective police force is also critical to the success of a counterinsurgency because there are limits to the use of military force—the enemy frequently holds little territory and refuses to fight for the territory that it does hold. Counterinsurgent military forces may be able to penetrate and garrison an insurgent area and, if well sustained, may reduce guerrilla activity. But, once the situation in an area becomes untenable for insurgents, they will simply transfer their activity to another area and the problem remains unresolved. As David Galula argues, "[C]onventional operations by themselves have at best no more effect than a fly swatter. Some guerrillas are bound to be caught, but new recruits will replace them as fast as they are lost."[21] A viable indigenous police force with a permanent presence in urban and rural areas is a critical component of counterinsurgency.

This poses a challenge for the U.S. military, since it is not ipso facto the lead U.S. agency for police training abroad. In the early 1970s, the U.S. Congress became deeply concerned that U.S. assistance abroad frequently strengthened the recipient governments' capacity for repression.[22] Consequently, Congress adopted Section 660 of the Foreign Assistance Act in 1974, which prohibited the United States from providing internal security assistance to foreign governments, stating that the U.S. government cannot

> provide training or advice, or provide any financial support, for police, prisons, or other law enforcement forces for any foreign government or any program of internal intelligence or surveil-

---

[20] Trinquier, *Modern Warfare*, p. 43; Galula, *Counterinsurgency Warfare*, p. 31.

[21] Galula, *Counterinsurgency Warfare*, p. 72.

[22] Michael McClintock, *The American Connection* (London: Zed Books, 1985); Martha K. Huggins, *Political Policing: The United States and Latin America* (Durham, N.C.: Duke University Press, 1998).

lance on behalf of any foreign government within the United States or abroad.[23]

The end of the Cold War and the increasing tempo of U.S. stability operations after 1989 rendered the 1974 legislation largely obsolete. U.S. government agencies increasingly secured waivers and provided police assistance to a range of regimes. Most foreign police training has been done by the U.S. Department of Justice's International Criminal Investigative Training Assistance Program, as well as the U.S. Department of State's Bureau of International Narcotics and Law Enforcement. Historically, the U.S. military provided limited police training in the context of stability operations. In Afghanistan and Iraq, however, the U.S. military greatly increased its police training initiatives.

Governance Capacity

There is also evidence that indigenous governance capacity impacts the outcome of counterinsurgencies.[24] The stronger and more competent a government is in providing services to its population, the greater its ability to undermine popular support for insurgents and the more likely that it can defeat an insurgency. Governance involves the provision of essential services to the population by a legitimate central authority in a timely manner. This provision of services can be impacted by such factors as the level of corruption, the viability of the justice system, and the influence of warlords and tribal militias.[25]

---

[23] U.S. House of Representatives, Committee on International Relations and U.S. Senate, Committee on Foreign Relations, *Legislation on Foreign Relations Through 2000* (Washington, D.C.: U.S. Government Printing Office, 2001), pp. 338–339; Robert M. Perito, *The American Experience with Police in Peace Operations* (Clementsport, Canada: The Canadian Peacekeeping Press, 2002), pp. 18–19.

[24] Ann Hironaka, *Neverending Wars: The International Community, Weak States, and the Perpetuation of Civil War* (Cambridge, Mass.: Harvard University Press, 2005); Fearon and Laitin, "Ethnicity, Insurgency, and Civil War," pp. 75–90. On the importance of building institutions, see Roland Paris, *At War's End: Building Peace After Civil Conflict* (New York: Cambridge University Press, 2004).

[25] On governance, see Daniel Kaufmann, Aart Kraay, and Massimo Mastruzzi, *Governance Matters III: Governance Indicators for 1996–2002* (Washington, D.C.: World Bank, 2002); Daniel Kaufmann, "Myths and Realities of Governance and Corruption," in *Global Com-*

The absence of good governance is often a root cause of an insurgency. A basic need of any insurgent group is an attractive cause. As David Galula argues, "The best cause for the insurgent's purpose is one that, by definition, can attract the largest number of supporters and repel the minimum of opponents."[26] All types of problems have been taken advantage of by insurgents. Sometimes the cause is social, such as the exploitation of one class by another. Class exploitation motivated a number of Marxist-Leninist insurgencies in Latin America, Africa, and Asia during the Cold War. Sometimes it is economic. The Chinese Communists capitalized on the plight of Chinese farmers, who were victims of exactions by authorities and the rapacity of local usurers. Poor governance may indicate disorganization, weakness, or incompetence—creating a window of opportunity for insurgents to win popular support.[27]

Corruption can be a particularly invidious challenge. It can undermine support for the government and increase support for insurgents. Corruption hampers economic growth, disproportionately burdens the poor, undermines the rule of law, and damages government legitimacy. It has a supply side (those who give bribes) and a demand side (public officials who take them).[28] At its core, corruption is the misuse of entrusted power for private gain. It can involve high-level officials with discretionary authority over government policies or lower-level officials who make decisions about enforcing (or not enforcing) regulations. Corruption also slows economic growth. It is often responsible for funneling scarce public resources away from projects that benefit the society and toward projects that benefit specific individuals. It hinders the development of markets and distorts competition, thereby

petitiveness Report 2005–2006 (Geneva: World Economic Forum, 2005), pp. 81–98; Paris, *At War's End*.

[26] Galula, *Counterinsurgency Warfare*, pp. 19–20.

[27] World Bank, *Reforming Public Institutions and Strengthening Governance* (Washington, D.C.: World Bank, 2000); Jessica Einhorn, "The World Bank's Mission Creep," *Foreign Affairs*, Vol. 80, No. 5 (2001), pp. 22–35.

[28] Transparency International, *Global Corruption Report 2006* (Berlin: Transparency International, 2006).

deterring investment. However, the most damaging effect of corruption is its impact on the social fabric of society: corruption undermines the population's trust in the political system, political institutions, and political leadership.[29]

In short, poor governance capacity contributes to declining support for the government, which can be fatal to counterinsurgencies. As Figure 2.4 shows, governments with high popularity defeated most of the insurgencies they fought, while unpopular governments lost more than one-half of the time.[30]

Figure 2.4
Government Popularity and Success

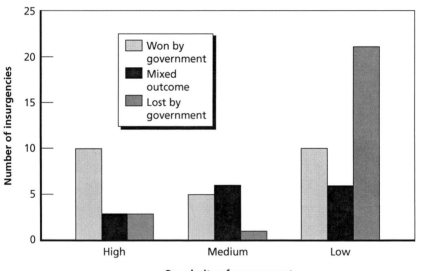

RAND MG595-2.4

---

[29] Kaufmann, "Myths and Realities of Governance and Corruption," pp. 81–98.

[30] We assessed the popularity of the government by making a qualitative judgment about how popular it was among the local population. We tried to avoid as much as possible the endogeneity problem of coding the government as popular if it won—and unpopular if it lost. Rather, we relied on the judgments of area specialists and historians that covered each insurgency to determine how popular the government was independent of the insurgency,

External Support

The final factor that impacts a counterinsurgency's success is external support to insurgents. The ability of insurgent groups to gain external support is correlated with their success. As Figure 2.5 illustrates, those insurgencies that received support from external states won more than 50 percent of the time, those with support from nonstate actors and diaspora groups won just over 30 percent of the time, and those with no external support won only 17 percent of the time.[31] Support from state actors and nonstate actors, such as a diaspora population, criminal network, or terrorist network, clearly makes a difference. This is intutive: Outside

Figure 2.5
External Support for Insurgents and Success

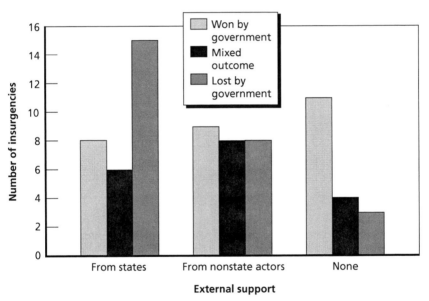

RAND MG595-2.5

---

which included examining public opinion polls where available. Unpublished RAND research for the U.S. Department of Defense.

[31] We assessed outside support from either state or nonstate actors using a dichotomous variable: yes if they received significant support from external actors and no if they received little or no support. Unpublished RAND research for the U.S. Department of Defense.

assistance—especially from states, which tend to have greater resources—can significantly bolster the capabilities of insurgent groups by giving them more money, weapons, logistics, and other aid.

External support can take two forms. First, foreign governments, diasporas, or international networks can provide direct assistance: training, operations, money, arms, logistics, diplomatic backing, and other types of aid.[32] The rise of a transnational jihadist network has created particularly acute challenges as organizations such as al Qaeda tap into local groups. These challenges include the flow of tactics, weapons, fighters, ideology, organization, and leadership into and among local insurgencies; the increased resort to suicide operations; and the pivotal role of religious figures in either fanning or opposing violence. Given the basic constraints posed by numerical weakness, insurgents need arms and materiel, money to buy them, or goods to trade for them. They need a supply of recruits, and they may also need information and instruction in the practical details of waging an insurgency.

The second type of external support is the freedom to use foreign territory as a sanctuary. This is sometimes made more tenable by the presence of a weak government where there is a sanctuary. The availability of a territorial base for insurgents outside of their home state is correlated with the failure of counterinsurgency efforts.[33] As Figure 2.6 illustrates, external sanctuary is a significant help to insurgents, often making the difference between their success or failure.[34] Insurgents have been successful approximately 43 percent of the time when they enjoyed a sanctuary.

While success in counterinsurgency warfare ultimately hinges on the ability to work with the indigenous government and its security forces, there are often significant challenges in doing so. The U.S. military faces at least two challenges in this area. First, the security forces

---

[32] Daniel L. Byman, *Deadly Connections: States That Sponsor Terrorism* (New York: Cambridge University Press, 2005), pp. 53–78.

[33] Fearon and Laitin, "Ethnicity, Insurgency, and Civil War."

[34] We assessed sanctuary using a dichotomous variable: yes if insurgents enjoyed significant sanctuary in a neighboring state or no if they had little or no sanctuary. Unpublished RAND research for the U.S. Department of Defense.

Figure 2.6
Sanctuary of Insurgents and Success

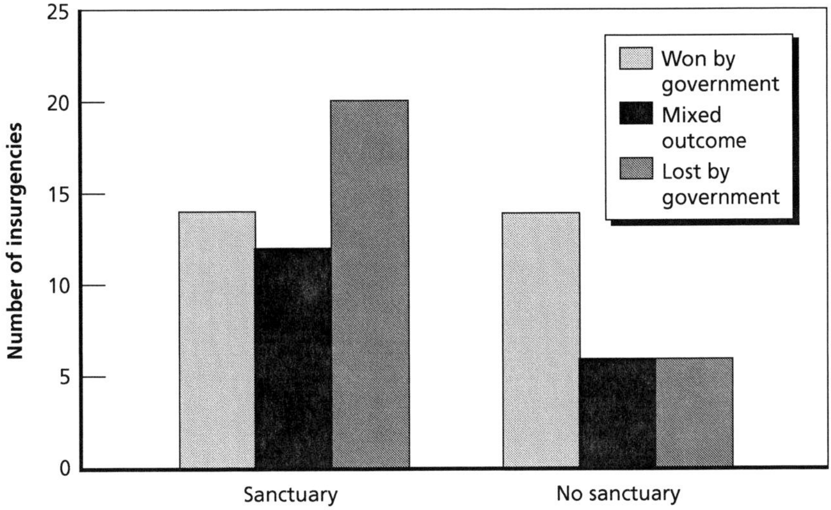

RAND *MG595-2.6*

may be poorly trained or corrupt. As Daniel Byman argues, many U.S. counterinsurgency allies have been characterized by poor intelligence, a lack of initiative, little integration of forces across units, soldiers who do not want to fight, bad leadership, and problems with training and creativity.[35]

Second, the indigenous government may be illegitimate, incompetent in providing basic services, weak, or corrupt. At the outset of the Afghan insurgency in 2002, for example, the central government was extremely weak. Afghan territory has historically been controlled by tribes and warlords, and its inhabitants have generally pledged loyalty to those with similar kinship ties and patrilineal descent rather than to the state authority.[36] Governance problems such as this can often be difficult to fix, especially in the near term. International assistance can build strong institutions in certain areas, such as central banking, which are isolated from society and responsive to the application

---

[35] Byman, *Going to War with the Allies You Have.*

[36] Rubin, *The Fragmentation of Afghanistan*, pp. 48–52.

of external technocratic expertise. But other areas, such as the justice system, are difficult to strengthen through the application of external assistance. These institutions have high "transaction volumes," are more deeply embedded in the social fabric, are an important element of the state's basis of legitimacy, and are heavily influenced by the cultural norms and values that shape institutions in any society.[37] Consequently, establishing a viable rule of law is an enormous challenge that can take a long time to accomplish.

Conclusion

External actors can play an important role in insurgencies and counterinsurgencies by tipping the balance in favor of either insurgents or the indigenous government. However, they usually cannot win it for either side, since locals have to govern and establish order over the long run. The indigenous force should be the default force of choice. Even if tactically successful, a unilateral operation by external forces may ultimately lead to failure by undermining and delegitimizing the very indigenous capability the external actor is trying to build.[38] Consequently, when the United States is involved in counterinsurgency warfare, the primary focus of its efforts should be to improve the performance and legitimacy of indigenous actors. This includes improving the quality of the police and other security forces, strengthening governance capacity, and undermining external support for insurgents. The rest of this study will further explore this argument.

---

[37] Francis Fukuyama, *State-Building: Governance and World Order in the 21st Century* (Ithaca, N.Y.: Cornell University Press, 2004).

[38] Walter Herd, *World War III: The Global Unconventional War on Terror* (Fort Bragg, N.C.: United States Army Special Operations Command, 2005).

CHAPTER THREE

The Age of Insurgency

Insurgencies are not new to Afghanistan. This chapter briefly examines Afghanistan's recent history of insurgency and argues that governance, the capacity of indigenous security forces, and external support have been critical factors in the outcome of these insurgencies. This finding has significant implications for understanding the resurgence of the Taliban that began in 2002.

In 1973, the royal dynasty that had ruled Afghanistan for more than two centuries fell. Mohammed Daoud deposed his brother-in-law, King Zahir Shah, and declared Afghanistan a republic. Daoud became president, abolished the monarchy, and forced Zahir Shah into exile in Rome. Marxist army officers helped consolidate Daoud's position, although this process was hampered by splits between the two main communist factions in Afghanistan: Khalq [the masses] and Parcham [the flag]. In 1978, Khalq army officers engineered a bloody coup, which led to the death of Daoud and his replacement by Nur Mohammad Taraki. Violence between the rival factions—including the murder of Taraki—coincided with wider rural revolts by Islamist opponents of the communist regime. Moscow grew increasingly concerned about the deteriorating security situation and feared that Taraki's successor, Hafizullah Amin, would turn to the West for assistance. Thus began Afghanistan's age of insurgency.

A Legacy of War

Over the next three decades, Afghanistan experienced at least four major insurgencies: the mujahideen wars against the Soviet Union (1979–1994), the rise of the Taliban (1994–2001), the U.S.-backed overthrow of the Taliban regime (2001–2002), and the return of the Taliban (2002–present). The objective of each of these insurgencies was to overthrow the existing regime and replace it with one more palatable to insurgent forces and their state sponsors. Many of the insurgent leaders—including Gulbuddin Hekmatyar, Abdul Rashid Dostum, Jalaluddin Haqqani, and Mullah Mohammed Omar—played key roles in most or all of the insurgencies.

The Mujahideen Period

In December 1979, the Soviet Union invaded Afghanistan, overthrew the Afghan government, and installed Babrak Karmal as leader. But a disparate collection of mujahideen insurgent groups resisted the Soviet occupation. The Soviets were successful in controlling the major cities and provincial towns in the country, but they never managed to take control of the countryside. Indeed, the situation in rural areas worsened for the Soviets and the Afghan government as mujahideen forces steadily gained popular support. Support for the mujahideen from Pakistan was a critical variable. Pakistan's Inter-Services Intelligence Directorate (ISI) provided money, weapons, training, and other assistance to Afghan insurgent groups, funneling aid to the mujahideen from a variety of other countries such as the United States and Saudi Arabia. Pakistan also provided sanctuary to mujahideen groups, where they were generally safe from Soviet forces.[1]

    Soviet losses mounted steadily, despite the Soviets' repeated efforts to defeat the mujahideen through the widespread deployment of mines, carpet-bombing of rebel areas, and the use of scorched-earth tactics.[2]

---

[1]   Steve Coll, *Ghost Wars: The Secret History of the CIA, Afghanistan, and bin Laden, from the Soviet Invasion to September 10, 2001* (New York: Penguin Books, 2004).

[2]   Lester Grau, ed., *The Bear Went Over the Mountain: Soviet Combat Tactics in Afghanistan* (Washington, D.C.: National Defense University Press, 1996); Lester Grau, *Artillery and*

In November 1986, Mohammad Najibullah was elected president of Afghanistan. He attempted to introduce a "national reconciliation" program, but with little success. When the Soviets withdrew in February 1989, the country was devastated. An estimated 1 million Afghans had been killed, more than 5 million had fled abroad, and 2–3 million were internally displaced. Nearly 15,000 Soviet soldiers had been killed, and as many as 500,000 had become sick or were wounded.[3] The Soviet withdrawal raised hopes both within Afghanistan and abroad for an imminent end to the conflict. However, fighting continued as the former anti-Soviet mujahideen coalition splintered along ethnic and political lines into competing factions. As a result of this in-fighting among the mujahideen forces, the pro-Moscow regime of President Najibullah was able to cling to power for three years after the Soviet withdrawal. In April 1992, Kabul finally fell to elements of the mujahideen, who then established a new government for the renamed the Islamic Republic of Afghanistan. Mujahideen leaders agreed to introduce a rotating presidency, starting with Burhanuddin Rabbani. However, disputes broke out over the division of government posts, and the fighting flared again. Pashtun leaders were particularly concerned about the makeup of the government and resented having to hand power over to other ethnic groups after more than 250 years of uninterrupted Pashtun rule.[4]

The Rise of the Taliban

By 1994, Afghanistan had disintegrated into a patchwork of competing groups and shifting alliances. The predominantly ethnic Tajik government of President Rabbani held Kabul and the northeast of the country, while the northern provinces remained under the control of

---

Counterinsurgency: The Soviet Experience in Afghanistan (Fort Leavenworth, Kan.: Foreign Military Studies Office, 1997).

[3] Ahmed Rashid, *Taliban: Militant Islam, Oil and Fundamentalism in Central Asia* (New Haven, Conn.: Yale University Press, 2000), p. 13; Barnett R. Rubin, *The Search for Peace in Afghanistan: From Buffer State to Failed State* (New Haven, Conn.: Yale University Press, 1995), p. 7; Grau, *The Bear Went Over the Mountain*, p. xix.

[4] On the 1988 Geneva Accords, which failed to establish peace in Afghanistan, see Rubin, *The Search for Peace in Afghanistan*.

Abdul Rashid Dostum and other warlords. Ismail Khan controlled the western provinces around Herat, and the area to the south and east of Kabul were in the hands of warlords such as Gulbuddin Hekmatyar. The eastern border with Pakistan was held by a council of mujahideen, and the south was split between scores of ex-mujahideen and bandits, who used their control of the roads to extort money from the cross-border trade with Pakistan.

In late 1994, a new movement emerged in the south, seizing control first of Kandahar and then of the surrounding provinces. Its leaders took the name of their group, Taliban, from the plural of talib, an Arabic word denoting an Islamic student. Many members were drawn from madrassas (Islamic theology schools) that had been established in Afghan refugee camps in northeastern Pakistan during the 1980s. The Taliban leadership, headed by Mullah Omar, presented itself as a cleansing force that would rid the country of the factionalism, corruption, and violence that had predominated since the Soviet withdrawal. Due to frustration and war-weariness among the population in the south, the Taliban was initially well received. Its forces advanced rapidly through southern and eastern Afghanistan, capturing nine out of thirty provinces by February 1995. The movement received strong backing from Pakistan's ISI, which assisted in the recruitment of members and provided weapons, training, and technical assistance.[5] In 1996, the Taliban captured Kabul and, despite temporary setbacks, conquered the northern cities of Mazar, Kunduz, and Taloqan in 1998.

By 2001, the Taliban controlled virtually all of Afghanistan. The only exception was a small sliver of land northeast of Kabul in the Panshjir Valley to which Ahmed Shah Massoud and his Northern Alliance forces had retreated. The Taliban instituted a repressive version of *shari'a* law, banning music, banned women from working or going to school, and prohibited freedom of the press. Afghanistan also became a breeding ground for jihadists and terrorists intent on attacking the United States and other nations. Osama bin Laden and his al Qaeda network used their money and influence to support the Taliban regime

---

[5] Coll, *Ghost Wars*, pp. 291–296, 331–332, 348, 414, 458.

and, in return, received permission to train operatives and plan operations on Afghan soil.[6]

Operation Enduring Freedom

After the September 11, 2001, terrorist attacks in Washington, New York, and Pennsylvania, the United States military launched Operation Enduring Freedom, helping Northern Alliance forces led by Abdul Rashid Dostum, Atta Mohammad Nur, Mohammad Qasim Fahim, and other local commanders mount a successful insurgency against the Taliban.[7] Over the next several months, U.S. and Afghan forces conducted a series of offensive operations, such as Operation Anaconda in the Shah-i-kot valley of eastern Afghanistan, against Taliban and al Qaeda forces.[8] The result was that most Taliban, al Qaeda, and other foreign jihadists resettled across the border in Pakistan. Although fighting continued for several years, the United States and other international actors began to assist Afghanistan with reconstruction by December 2001.

As the Taliban's power base collapsed, international and local attention turned to nation-building. The UN helped organize a meeting of Afghan political leaders in Bonn, Germany, in late November 2001. On December 5, 2001, Afghan leaders signed the Bonn Agreement. It established a timetable for a transition to legitimate power structures, which culminated in the establishment of a fully representative and freely elected government. The UN Security Council endorsed

---

[6] *The 9/11 Commission Report: Final Report of the National Commission on Terrorist Attacks Upon the United States* (New York: W. W. Norton, 2004); Coll, *Ghost Wars*, pp. 327–344, 363–365, 379–386, 400–415.

[7] On the overthrow of the Taliban regime, see Schroen, *First In*; Biddle, *Afghanistan and the Future of Warfare*; Berntsen and Pezzullo, *Jawbreaker*; and Woodward, *Bush at War*.

[8] On Operation Anaconda, see U.S. Air Force, Office of Lessons Learned (AF/XOL), *Operation Anaconda: An Air Power Perspective* (Washington, D.C.: Headquarters United States Air Force AF/XOL, February 2005); Paul L. Hastert, "Operation Anaconda: Perception Meets Reality in the Hills of Afghanistan," *Studies in Conflict and Terrorism*, Vol. 28, No. 1, January–February 2005, pp. 11–20; and Sean Naylor, *Not a Good Day to Die: The Untold Story of Operation Anaconda* (New York: Berkley Books, 2005).

the outcome the following day in Resolution 1383.[9] Under the Bonn Agreement, the parties agreed to establish an interim authority comprising three main bodies: a 30-member interim administration headed by Hamid Karzai, a Pashtun, which took power on December 22; a supreme court; and a Special Independent Commission for the Convening of the Emergency Loya Jirga.[10]

In January 2002 in Tokyo, international donors pledged over $4.5 billion for reconstruction efforts. The parties present at Bonn had also asked the United Nations to "monitor and assist in the implementation of all aspects" of the agreement.[11] To that end, Security Council Resolution 1401, passed on March 28, 2002, established the UN Assistance Mission in Afghanistan (UNAMA). In addition, the United Kingdom agreed to be the lead nation for counternarcotics, Italy for justice, the United States for the army, Germany for police, and Japan for the disarmament, demobilization, and reintegration of former combatants. The Emergency Loya Jirga, which was attended by approximately 2,000 people, took place between June 12 and 19, 2002, following extensive preparations and countrywide consultations. At the conclusion, Hamid Karzai was chosen as president of the transitional administration and head of state. His nominees for key posts in the administration were also approved by the Emergency Loya Jirga. The defense and foreign affairs portfolios were given to the mainly Tajik Northern Alliance, while the Ministry of Interior went to a Pashtun regional governor.

The Return of the Taliban
In the spring and summer of 2002, Taliban and other forces began to conduct offensive operations to overthrow the Afghan government and coerce the withdrawal of U.S. and coalition forces.[12] In April, for exam-

---

[9] *United Nations Security Council Resolution 1383*, S/RES/1383, December 6, 2001.

[10] A *loya jirga* is a traditional meeting of Afghan tribal, political, and religious leaders.

[11] United Nations Assistance Mission in Afghanistan, "Agreement on Provisional Arrangements in Afghanistan Pending the Reestablishment of Permanent Government Institutions," December 2001.

[12] Some have argued that the insurgency began in earnest in June 2004. But Taliban offensive operations two years earlier suggest that it was in the spring of 2002. COL Walter M.

ple, Taliban and jihadist forces conducted a series of offensive attacks in Kandahar, Khowst, Nangarhar, Kabul, and other Afghan provinces. This marked the beginning of the latest insurgency. After 2002, insurgent groups waged an increasingly violent campaign in Afghanistan despite political progress. In January 2004, for example, Afghans adopted a new constitution. In October 2004, they elected Hamid Karzai as president, despite efforts by the Taliban and other insurgent groups to target those involved in the election. In September 2005, Afghans elected a new parliament, which included a number of ex-Taliban ministers. At least several individuals formerly associated with the Taliban won Wolesi Jirga (lower house of the National Assembly) seats in the September 2005 elections, including Abdul Salam Rocketi. President Karzai appointed the former Taliban deputy religious affairs minister, Mawlawi Arsallah Rahmani, to the Meshrano Jirga (upper house) along with Gulbuddin Hekmatyar's former close ally, Abdul Saboor Farid.[13] Despite these steps, however, the insurgency continued to worsen.

Key Themes

This brief overview of Afghanistan's age of insurgency highlights three factors that have contributed to the success of past insurgencies: governance challenges, external support, and variations in the quality of security forces.

Governance

Afghanistan has a long history of decentralized governance. Following the second Anglo-Afghan war in 1880, Amir Abdul Rahman Khan seized power after the departure of British troops. With British finan-

---

Herd, COL Patrick M. Higgins, LT COL Adrian T. Bogart, III, MAJ A. Davey, and CAPT Daudshah S. Andish, *One Valley at a Time* (Fort Bragg, N.C.: Combined Joint Special Operations Task Force–Afghanistan, 2005), p. 121.

[13] See, for example, International Crisis Group, *Countering Afghanistan's Insurgency: No Quick Fixes* (Kabul: International Crisis Group, 2006).

cial and military assistance, he ruthlessly attempted to defeat or manipulate tribal and ethnic groups such as the Hazaras, Aimaqs, Nuristanis, and various Pashtun tribal coalitions. However, Khan was unable to destroy tribal power and establish a strong, centrally controlled state.[14] Successive efforts over the next century generally failed. Afghan territory has been controlled by tribes and local strongmen, and its inhabitants have generally pledged loyalty to those with similar kinship ties and patrilineal descent rather than to a central governing authority.[15] One consequence of this power structure is that Afghan governments have never been able to establish a monopoly on the legitimate use of force inside the country. The insurgencies that began with the 1979 Soviet invasion and continued through the Taliban conquests in the 1990s only served to increase Afghanistan's decentralized political structure. In addition, Afghan governments have never established a formal justice system. In the absence of a central government, local shuras (village councils) and tribal elders developed an informal legal system that incorporated a wide range of Islamic and customary laws.[16]

External Support

State support and sanctuary have been critical variables in the outcome of these insurgencies. Afghan governments and opposition groups have received aid from a number of states. The Soviet Union provided a total of $1.3 billion in economic aid and $1.3 billion in military aid to the Afghan government between 1955 and 1978 and roughly $5 billion per year between 1979 and 1989. The United States provided $533 million in economic aid to the Afghan government between 1955 and 1978, and between $4 billion and $5 billion to the mujahideen between 1980

---

[14] Rubin, *The Fragmentation of Afghanistan*, pp. 48–52.

[15] Richard Tapper, "Anthropologists, Historians, and Tribespeople on Tribe and State Formation in the Middle East," in Philip S. Khoury and Joseph Kostiner, eds., *Tribes and State Formation in the Middle East* (Berkeley, Calif.: University of California Press, 1990).

[16] Olivier Roy, *Islam and Resistance in Afghanistan,* 2nd ed. (New York: Cambridge University Press, 1990); Rashid, *Taliban*, pp. 1–13; Barnett R. Rubin, "(Re)Building Afghanistan," *Current History*, Vol. 103, No. 672 (April 2004), pp. 165–170.

and 1992.[17] Both the United States and the Soviet Union suspended most aid in 1991. The Pakistan government, especially the ISI, played a particularly active role in Afghan politics. Pakistan provided significant assistance to the mujahideen during the Soviet wars. And it provided weapons, financial aid, and other assistance such as wheat and petroleum to the Taliban and other groups from the 1990s through 2001.[18] Saudi Arabia gave nearly $4 billion in official aid to the mujahideen between 1980 and 1990; there was also a flow of unofficial aid from Saudi Islamic charities and foundations, the private funds of Saudi princes, and mosque collections. Saudi Arabia provided aid to the Taliban and al Qaeda in Afghanistan until 1998.[19] Finally, Iran provided assistance to various factions, especially to Afghan commanders in the western regions of the country. Iranian military aid to the anti-Taliban alliance escalated after the fall of Kabul in 1996 and again after the fall of Mazar in 1998.

Security Forces
Afghanistan's history of weak central governments and the flow of support from external actors impacted efforts to establish strong central government forces. During the 1980s and 1990s, there was no national civilian police force in Afghanistan. Instead, local militia and tribal forces enforced the rule of law in much of the country. Among the Pashtun (who constitute a majority of Afghanistan's population), the traditional military institution has been the lashkar, or the armed tribe. Each family contributed male members and weapons to the lashkar, which was further enriched by whatever material it took in battle.

---

[17] Rashid, *Taliban*, p. 18; Rubin, *The Fragmentation of Afghanistan*, p. 20.

[18] Tim Judah, "The Taliban Papers," *Survival*, Vol. 44, No. 1 (Spring 2002), pp. 69–80; Rubin, *The Fragmentation of Afghanistan*, pp. 196–225; Ahmed Rashid, "Pakistan and the Taliban," in William Maley, ed., *Fundamentalism Reborn? Afghanistan and the Taliban* (New York: New York University Press, 2001), pp. 72–89.

[19] Samuel P. Huntington, *The Clash of Civilizations and the Remaking of World Order* (New York: Simon and Schuster, 1996), pp. 246–248; *The 9/11 Commission Report*, pp. 63–67, 371–374; Rashid, *Taliban*, pp. 48, 54; Stockholm International Peace Research Institute, *SIPRI Yearbook 1991: World Armaments and Disarmament* (New York: Oxford University Press, 1991), p. 199.

There are no precise figures for the number of militia fighters. Estimates during the 1990s ranged from about 200,000 to 600,000.[20] Most were untrained, ill-equipped, illiterate, and owed their allegiance to local warlords and military commanders, not to the central government.[21] As a German delegation in January 2002 concluded

> The police force is in a deplorable state just a few months after the dissolution of the Taliban regime. There is a total lack of equipment and supplies. No systematic training has been provided for about 20 years. At least one entire generation of trained police officers is missing. Next to constables, former Northern Alliance fighters are being put to work as police officers.[22]

While Afghanistan has lacked a trained civilian police force, it has had secret police agencies. During the Soviet era, the Afghan government established a powerful secret police body, the State Information Services, to suppress opponents of the regime and establish order. The Taliban established the Ministry of Enforcement of Virtue and Suppression of Vice to enforce decrees regarding moral behavior, such as those restricting women's employment, education, and dress; enforcing men's beard length and mosque attendance; and regulating the activities of the United Nations and NGOs.[23]

The Afghan army has traditionally had little internal control over the country. During the 1980s, the Moscow-backed Democratic Republic of Afghanistan and its successor, the Republic of Afghanistan, recruited a large number of tribal and local militias as army forces.

---

[20] Ali A. Jalali, "Afghanistan: The Anatomy of an Ongoing Conflict," *Parameters*, Vol. XXXI, No. 1 (Spring 2001), p. 86; International Institute for Strategic Studies, *The Military Balance, 1995/96* (London: Oxford University Press, 1995), pp. 155–156; International Institute for Strategic Studies, *The Military Balance, 1998/99* (London: Oxford University Press, 1998), pp. 153–154.

[21] Laurel Miller and Robert Perito, *Establishing the Rule of Law in Afghanistan*, Special Report 117 (Washington, D.C.: United States Institute of Peace, 2004).

[22] Government of Germany, Federal Foreign Office and Federal Ministry of Interior, *Assistance in Rebuilding the Police Force in Afghanistan* (Berlin: Federal Foreign Office and Federal Ministry of the Interior, March 2004), p. 6.

[23] Rubin, *The Fragmentation of Afghanistan*, p. xv; Rashid, *Taliban*, p. 106.

During the Taliban era, the army was comprised of an assortment of armed groups with varying degrees of loyalties and professional skills. There was no formal military structure. The army was not organized, armed, or commanded by the state.[24] Mullah Mohammed Omar was commander of the armed forces and ultimately decided on military strategies, key appointments, and military budgets. A military shura sat below Omar, helping to plan strategy and implement tactical decisions. Individual Taliban commanders were responsible for recruiting men, paying them, and looking after their needs in the field. These field commanders acquired much of the money, fuel, food, transport, and weapons they needed from the military shura. The Taliban's military structure also included Pakistani officers and al Qaeda members. For example, the elite Brigade 055 consisted of Pakistani, Sudanese, and other foreign fighters.[25]

Conclusion

This overview reiterates the argument laid out in Chapter Two: Governance challenges, external support, and the capacity of indigenous security forces have been critical factors in the outcome of Afghanistan's insurgencies. Pakistan played two particularly important roles. First, the Pakistan government, especially the ISI, supported the victors of each insurgency: the mujahideen, Taliban, and U.S. forces during the initial stages of Operation Enduring Freedom. Since Afghanistan and Pakistan share a 1,160-mile border, Pakistani leaders have historically viewed the ability to influence Afghanistan as critical for strategic depth. Second, Afghan insurgent groups have repeatedly used Pakistan as a sanctuary. These findings have significant implications for understanding the Taliban's resurgence and assessing how they can be defeated. The next three chapters will explore these findings and their implications by focusing on the three key actors: insurgents, the

---

[24] Jalali, "Afghanistan."

[25] Rashid, *Taliban*, pp. 98–100.

Afghan government and its security forces, and the United States and other coalition partners.

CHAPTER FOUR

Insurgents and Their Support Network

This chapter examines Afghan insurgent groups and their support network in the early stages of the insurgency. It argues that a critical independent variable in the success of any counterinsurgency is outside support for insurgents. The insurgency in Afghanistan included a dangerous combination of local and transnational support. Afghan groups successfully acquired external support and assistance from the global jihadist network, including groups with a strong foothold in Pakistan, such as al Qaeda. They also acquired support from some individuals in the Pakistan government, as well as local tribes, criminal organizations, and militias in Pakistan and Afghanistan. This assistance enabled Afghan insurgent groups to adapt their tactics, techniques, and procedures—to become, in effect, learning organizations—and largely explains the resurgence of the Taliban and other insurgent forces.

This chapter is divided into four sections. The first outlines the main insurgent groups. The second examines their increasing ability to conduct violence. The third section assesses why they have been successful in destabilizing the south and east, especially their ability to gain external support. The fourth offers a brief conclusion.

Insurgent Groups

The insurgency in Afghanistan included six main insurgent groups: the Taliban, Hezb-i-Islami, the Haqqani network, foreign fighters (mostly Arabs and Central Asians), tribes based in Pakistan and Afghanistan, and criminal networks. There is evidence of some coordination among these groups at the tactical, operational, and strategic levels—including through several shuras located in Pakistan.[1] But there was no unified leadership. As Figure 4.1 illustrates, Afghan insurgent groups fell into three loose fronts. The northern front, which included a large Hezb-i-Islami presence, was based in Afghan provinces that include Nurestan, Konar, Laghman, and Nangarhar, as well as across the border in Pakistan. Other groups, such as Lashkar-e-Taiba and Tehreek-e-Nefaz-e-Shariat-Mohammadi, were also active on this front. The central front included a loose amalgam of foreign fighters, including Central Asians and Arabs. They were located in a swath of territory near the Afghanistan-Pakistan border from Bajaur in the Federally Administered Tribal Areas in Pakistan to Khowst, Paktia, and Paktika in Afghanistan. The Haqqani network was active in the central front against Afghan and coalition forces. Finally, the southern front, which included a large Taliban presence, was based in Baluchistan and the Federally Administered Tribal Areas of Pakistan, as well around the Afghan provinces of Helmand, Kandahar, Oruzgan, Zabol, and Paktika. In addition to the Taliban, several drug and tribal groups were also active in the southern front.

The Taliban

The Taliban have historically been motivated to impose a radical interpretation of Sunni Islam in Afghanistan, which is derived from the Deobandi school of thought.[2] Their primary strategies for accomplishing this objective have long been to overthrow the Afghan government, break the political will of the United States and its coalition partners,

---

[1] On cooperation among insurgents, see Barnett R. Rubin, *Afghanistan and the International Community: Implementing the Afghanistan Compact* (New York: Council on Foreign Relations, 2006); "Afghan Taliban Say No Talks Held with U.S., No Differences with Hekmatyar," *Karachi Islam*, February 24, 2005, pp. 1, 6; "Pajhwok News Describes Video of Afghan Beheading by 'Masked Arabs,' Taliban," Pajhwok Afghan News, October 9, 2005; "Spokesman Says Taliban 'Fully Organized,'" *Islamabad Ausaf*, June 23, 2005, pp. 1, 6; "UK Source in Afghanistan Says al Qaeda Attacks Boost Fear of Taliban Resurgence," *Guardian* (London), June 20, 2005; "Taliban Military Chief Threatens to Kill U.S. Captives, Views Recent Attacks, Al-Qa'ida," interview with Al Jazeera TV, July 18, 2005.

[2] Roy, *Islam and Resistance in Afghanistan*; Rashid, *Taliban*; William Maley, ed., *Fundamentalism Reborn? Afghanistan and the Taliban*, New York: New York University Press, 2001.

Figure 4.1
The Afghan Insurgent Front

and coerce foreign forces to withdraw. As former Taliban spokesman
Mofti Latifollah Hakimi argued, "the only avenue open to us is the
path of jihad."[3] Indeed, the Taliban increasingly adopted jihadist rhetoric to reestablish control of the country. The Taliban included a recent
influx of new members—sometimes referred to as the "neo-Taliban"—
who were recruited at madrassas and other locations in Afghanistan
and Pakistan. There were several thousand full-time Taliban fighters;
some estimates ranged from 5,000 to 10,000.[4]

---

[3] "Spokesman Rejects Afghan Government's Amnesty Offer for Taliban Leader," Peshawar
Afghan Islamic Press (May 9, 2005).

[4] Estimates of insurgents are notoriously difficult for two reasons. First, it is difficult to
count the number of insurgents since they hide in urban and rural areas to evade foreign and
domestic intelligence and security forces. Second, the number of insurgents is often fluid.
Some are full-time fighters, but many are not. In addition, there is a significant logistics,

The Taliban has traditionally involved two main tiers. The top tier included the leadership structure and key military and political guerrillas and commanders. They were motivated by a radical version of Islam and saw the insurgency as a fight between Islam and Western infidels and the West's "puppet government" in Kabul. The Taliban leadership was comprised of Mullah Omar and his senior lieutenants, many of whom were based in Quetta, Pakistan.[5] A second shura was based in Pakistan's Federally Administered Tribal Areas and revolved around such individuals as Sirajuddin Haqqani, who commanded several hundred fighters and was loosely allied with Taliban leadership.[6] His base of support was in the Afghan provinces of Khowst, Logar, Paktia, and Paktika, as well as in Waziristan.[7] Haqqani cooperated with the Pakistan government, including the military and ISI. He enjoyed a support base and ran madrassas in and around Miranshah and Mir Ali.[8] In addition, the Taliban organized a parallel Afghan government, which

---

financial, and political support network for insurgent groups, making it virtually impossible to reliably estimate the total number of guerrillas and their support base. These reasons make it more difficult to estimate the number of insurgents than to estimate the size of state military forces. On the Taliban numbers, the author interviewed U.S., European, and Afghan officials on numerous occasions throughout 2004, 2005, and 2006.

[5]   There is an extensive body of literature on the Taliban's base of operations in Pakistan; see, for example, Jason Burke, "The New Taliban," *Observer* (London), October 14, 2007, p. 31; Barnett R. Rubin, "Still Ours to Lose: Afghanistan on the Brink," submitted as written testimony to the Senate Foreign Relations Committee, September 21, 2006; and Barnett R. Rubin, "Saving Afghanistan," *Foreign Affairs*, Vol. 86, No. 1, January/February 2007.

[6]   David Rohde, "Foreign Fighters of Harsher Bent Bolster Taliban," *New York Times*, October 30, 2007, p. A1; Declan Walsh, "Pakistan: Resurgent al-Qaida Plotting Attacks on West From Tribal Sanctuary," *Guardian* (London), September 27, 2007, p. 24; and Burke, "The New Taliban."

[7]   Syed Saleem Shahzad, "Taliban's New Commander Ready for a Fight," *Asia Times* (May 20, 2006).

[8]   On Pakistan raids against Haqqani, see Iqbal Khattak, "40 Militants Killed in North Waziristan," *Daily Times* (Pakistan), September 30, 2005; "Pakistani Law Enforcers Intensify Hunt for Haqqani," Pajhwok Afghan News, March 7, 2006. On Haqqani's historical role also see Charles Dunbar, "Afghanistan in 1986: The Balance Endures," *Asian Survey*, Vol. 27, No. 2: pp. 127–142.

included governors for Afghan provinces and ministers for such areas as defense and justice.[9]

The bottom tier of Taliban guerrillas included thousands of local Afghan fighters. They were primarily men from rural villages who were paid to set up roadside bombs, launch rockets and mortars at NATO and Afghan forces, or pick up a gun for a few days. Most were not ideologically committed to jihad. Rather, they were motivated by unemployment, disenchantment with the lack of change since 2001, or anger over the killing or wounding of a local villager by Afghan, U.S., or NATO forces. Some fought because of grievances with the Afghan government or because of abuse (either actual or perceived) by Afghan or coalition forces—such as bombings and intrusive house searches.

Hezb-i-Islami

Gulbuddin Hekmatyar's Hezb-i-Islami, which included several hundred fighters, sought to overthrow the government of Afghanistan and install Hekmatyar as leader.[10] Hekmatyar is a Pashtun from the Imam Sahib district of Kunduz and served as Afghanistan's prime minister from March 1993 to 1994 and again briefly in 1996. His group has traditionally found support in the areas around the Afghan provinces of Konar, Nurestan, Nangarhar, Paktia, and Paktika.[11] The Hezb-i-Islami has historically received assistance from the Iran and Pakistan governments, as well as the United States during the Cold War.[12] Hekmatyar openly pledged to cooperate with al Qaeda and Taliban forces to fight the "crusader forces" in Afghanistan.[13]

Both the Taliban and Hezb-i-Islami had loosely hierarchical organizational structures and were roughly divided into four groups: logistics support, financial and political support, guerrillas, and com-

---

[9] Amrullah Saleh, *Strategy of Insurgents and Terrorists in Afghanistan* (Kabul: National Directorate for Security, 2006), p. 2.

[10] Author interview with U.S. government officials, Kabul, Afghanistan, September 2006.

[11] Saleh, *Strategy of Insurgents and Terrorists in Afghanistan*, p. 2.

[12] On Hekmatyar's relationship with Pakistan, see Coll, *Ghost Wars*, pp. 181–183, 210–211.

[13] See, for example, "Al Jazeera Airs Hikmatyar Video," Al Jazeera TV, May 4, 2006.

manders.[14] The logistics support network provided supplies, equipment, and other assistance.[15] The success or failure of the guerrilla force depends to a great extent on the logistics support network's ability to gain support from the tribal populations in Afghanistan and Pakistan. Individuals in this group assisted the insurgency by acquiring supplies (including food, water, and ammunition), conducting information and intelligence campaigns, operating medical facilities, conducting counterintelligence operations, recruiting new guerrillas or supporters, operating communication systems, and acquiring and maintaining equipment. The financial and political support network was focused on acquiring money and other types of assistance from government and nongovernment sources, and establishing political relations with friendly governments, tribal leaders, and other groups. It also worked to recruit new members from madrassas, tribes, and foreign countries. The guerrillas were the armed insurgents, who conducted military and paramilitary operations. Insurgent groups also paid locals—especially youths—to conduct low-level operations, such as harassing fire at U.S. or coalition forces. The Taliban often paid these local youths two or three times the daily rate of pay of ANA and ANP units, suggesting that some insurgents were motivated more by financial than ideological incentives.[16] Finally, the commanders provided strategic organization. Commanders did not exert control in traditional military terms. Rather, guerrilla units were often given tactical and operational autonomy.[17]

---

[14] Herd et al., *One Valley at a Time*, pp. 67–76.

[15] Key Afghan cities for the support network include those along the ring road, such as Kabul, Jalalabad, Ghazni, Gardez, Qalat, Kandahar, and Herat. In addition, the main auxiliary lines include the old trade routes and way stations that connect Afghanistan cities to Iran and Afghanistan. Examples include the old Silk Road to Tehran, Pepper Route through Peshawar and on to India, route through the Khyber Pass to Peshawar, and route from Spin Boldak to Quetta.

[16] In the summer of 2006, for example, U.S. and NATO military officials reported that the Taliban paid locals in some areas roughly $14 per day to fight, compared to approximately $4 for many ANA soldiers.

[17] Herd et al., *One Valley at a Time*, pp. 67–76. When the Taliban fell from power, Haqqani told local reporters,

Foreign Fighters

The foreign fighters assisting the Afghan insurgents were an amalgam of loosely knit Muslim extremists, including al Qaeda forces. Their ranks were made up primarily of two major types: individuals from the Caucasus and Central Asia (such as Chechens, Uzbeks, and Tajiks) and Arabs (such as Saudis, Egyptians, and Libyans). Several waves of Arabs settled in Waziristan during the mujahideen wars against the Soviets and, more recently, after the U.S. and Northern Alliance overthrow of the Taliban regime. A number of foreigners were directly or indirectly affiliated with al Qaeda, though some were simply inspired by the broader jihadist goal of pushing the United States and its Western allies out of Afghanistan.[18]

Al Qaeda played an important role as an enabler of the Taliban and had close relations with several Pakistan militant groups such as Jaish e Muhammad, Harakat ul Mujahideen, Lashkar e Jhangvi, and Harakat ul Jihad ul Islami. Key al Qaeda figures involved in the Afghanistan insurgency included

- Ayman al-Zawahiri
- Mustafa Abu al-Yazid
- Abu Yahya al-Libi
- Adam Gadahn.

Indeed, the Afghan-border region of Pakistan was home to what is generally referred to as "al Qaeda central"—the remnants of the

---

We will retreat to the mountains and begin a long guerrilla war to reclaim our pure land from infidels and free our country like we did against the Soviets . . . we are eagerly awaiting the American troops to land on our soil, where we will deal with them in our own way. (Scott MacDonald, "Minister's Visit Hints at Taliban Split," Reuters, October 20, 2001)

[18] On the role of Al Qaeda in Afghanistan, see J. Michael McConnell, *Annual Threat Assessment of the Director of National Intelligence for the Senate Select Committee on Intelligence,* statement for the record, February 5, 2008; National Intelligence Council, *The Terrorist Threat to the U.S. Homeland* (Washington, D.C.: National Intelligence Council, 2007); and John D. Negroponte, *Annual Threat Assessment of the Director of National Intelligence for the Senate Armed Services Committee,* statement for the record, February 28, 2006.

pre-9/11 al Qaeda organization.[19] Osama bin Laden and Zawahiri relied on an informal infrastructure of militants and tribes to survive, travel, communicate, and conduct operations. First forged during the Afghan jihad of the 1980s, these relationships were institutionalized in the late 1990s when al Qaeda trained and catalogued tens of thousands of Pakistani militants in camps inside Afghanistan. Al Qaeda was deeply involved in suicide terrorism, which will be discussed in more detail below. For example, Croma Yahya, a suicide bomber from Mali who failed to kill Balkh governor Atta Mohammad Nur, had al Qaeda links and came to Afghanistan through Pakistan.[20] As Lt. General Michael Maples, head of the Defense Intelligence Agency, argued, "Al Qaeda will remain engaged in Afghanistan for ideological and operational reasons. Taliban and other anti-coalition militants are adopting al Qaeda tactics in Afghanistan."[21]

The map in Figure 4.2 illustrates Pakistan's tribal areas, which stretch for 500 miles along the Afghanistan-Pakistan border. The foreign fighters often acted as franchises to al Qaeda. They had autonomy at the tactical and operational level but often took guidance from more senior al Qaeda or other commanders at the strategic level. Their strategic objectives were much broader than those embraced by Hezb-i-Islami and Taliban forces. They adopted a strict interpretation of Islam, embraced jihad against the U.S. and other allied governments, and sought to eradicate Western military forces and influ-

---

[19] Al Qaeda today can be conceptualized in several dimensions. The first includes al Qaeda affiliates and associates. These are terrorist groups that have directly benefited from Osama bin Laden's spiritual guidance and received substantial training, arms, money, and other assistance. The second includes amorphous groups of al Qaeda adherents, who may have received some training in al Qaeda facilities, some encouragement and direction, and perhaps a minimal amount of assistance. The third dimension includes homegrown Islamic radicals who have no direct connection with al Qaeda but who nonetheless are prepared to carry out attacks in solidarity with al Qaeda's radical jihadist agenda. The final dimension is al Qaeda central. See, for example, Bruce Hoffman, *Inside Terrorism* (New York: Columbia University Press, 2006), pp. 285–289.

[20] Author interview with officials from several Western, Afghan, and Pakistani government agencies, 2005 and 2006.

[21] LTG Michael D. Maples, *Current and Projected National Security Threats to the United States, Statement for the Record, Senate Armed Services Committee*, February 28, 2006.

Figure 4.2
Pakistan Tribal Areas

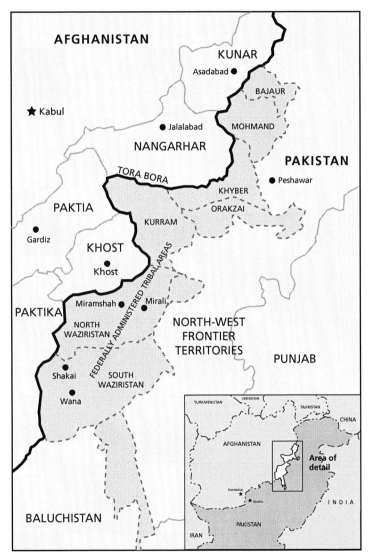

ence from the region.[22] However, their objectives were usually much broader than regime change in Afghanistan, and included a return of the Islamic caliphate in the Middle East. Foreign jihadists were often better equipped, trained, and motivated than other insurgent forces, and played a key role as trainers, shock troops, and surrogate leaders for Taliban units in the field.[23]

Tribes

There were a number of Afghan and Pakistan tribes—especially Pashtun tribes—that allied with insurgent groups and provided assistance to the insurgency. For example, the Taliban had a significant support network among the Ghilzai tribes, as well as among such Durrani tribes as the Norzais, Alekozai, and Eshaqzais. There were also a number of groups, such as Ahmadzai Wazirs and Mahsuds in North and South Waziristan, that coordinated activities for and provided support to insurgent groups. As one Afghan National Directorate for Security report argued,

> [The Taliban's use of] recruitment techniques in the ongoing stage is becoming sophisticated. They approach tribes, sub-tribes and communities in the villages. They want them to sever their relationship with the government and also preach to the population to support the jihad against the Americans and the government which they consider the infidel.[24]

Several individuals in the Ahmedzai Wazir tribe based in Wana, Pakistan, helped raise funds and recruited militants to fight in Afghanistan. There was also evidence that organizations such as Jamaat-e-Islami and

---

[22] On salafi jihadists, see Gilles Kepel, *Jihad: The Trail of Political Islam* (Cambridge, Mass.: Harvard University Press, 2002), pp. 219–222; Guilain Denoeux, "The Forgotten Swamp: Navigating Political Islam," *Middle East Policy*, Vol. IX, No. 2 (June 2002), pp. 69–71.

[23] There have been a handful of cases in which foreign fighters have charged fire bases and special forces camps in the open in virtual suicide missions. Some of these tactics have raised questions about the competence of at least some foreign fighters.

[24] Saleh, *Strategy of Insurgents and Terrorists in Afghanistan*, p. 3.

Wahhabi groups such as Lashkar-e-Taiba cooperated with insurgent groups.[25]

Criminal Groups

A variety of organizations involved in the illicit drug trade also cooperated with the Taliban, especially in such Afghan provinces as Helmand. In many cases, members of the Afghan government—including the ANP—were also involved in narcotics trafficking. Former Afghan Minister of Interior Ali Jalali argued that rising narcotics-related violence "is more indicative of a change in tactics than capability," including "closer cooperation between the militants and drug-traffickers (particularly in Helmand Province)."[26] Drug and other criminal groups developed an intricate network to transport drugs between Afghanistan and neighboring states. The Taliban profited from the drug trade in several ways. They received payments to provide protection to some drug-trafficking organizations operating in Afghanistan and Pakistan. The Taliban also levied taxes on some farmers and secured bribes from drug-trafficking groups at checkpoints.[27] Farther north, there were a number of Russian, Tajik, Uzbek, and Turkmen drug-trafficking organizations. Tajikistan has historically been a primary transshipment country for opiate shipments destined for Russia. Drug traffickers in Afghanistan used produce-laden trucks as a cover for drugs sent north toward Tajikistan, where they were handed off to other criminal organizations. Tajik criminal organizations were the primary movers of this contraband. Approximately half of the heroin that passed through

---

[25] On Ismail, see "Taliban Claim Shooting Down U.S. Helicopter," *The News* (Islamabad), June 29, 2005. On Wana, see Intikhab Amir, "Whose Writ Is It Anyway?" *The Herald* (Pakistan), April 2006: pp. 80–82.

[26] Ali A. Jalali, "The Future of Afghanistan," *Parameters*, Vol. XXXVI, No. 1 (Spring 2006), p. 8.

[27] Author interview with Western and Afghan government officials, 2005, 2006, and 2007; Karen P. Tandy, *Statement of Karen P. Tandy, Administrator, U.S. Drug Enforcement Agency, Testimony Before the House Armed Services Committee*, June 28, 2006.

Tajikistan was consumed in Russia. The balance transited Russia to other consumer markets in Western and Eastern Europe.[28]

In places where the local economy was dominated by drug traffickers, the opium poppy economy financed a mode of local government that undermined the power of the central government.[29] Researchers at the World Bank argued that areas of Afghanistan—particularly in the south—became a fragmented narco-state in which local drug lords took control of district- and provincial-level state institutions of government with the assistance of insurgent groups.[30] In short, the opium industry has become increasingly powerful and institutionalized in Afghanistan "as Afghan traffickers and the armed leaders who profit from them—both warlords within the government and anti-government forces—capture a higher proportion of the value added of the opiate trade."[31]

Insurgent Activity

From the beginning of the current Afghan insurgency in 2002, there was a gradual deterioration in the security environment—especially in the south and east of the country. RAND data show that the overall number of insurgent-initiated attacks increased approximately 400 percent from 2002 to 2006, and the number of deaths from these attacks increased over 800 percent during the same period.[32] The data

---

[28] Tandy, *Statement*.

[29] Jan Koehler, *Conflict Processing and the Opium Poppy Economy in Afghanistan* (Jalalabad: Project for Alternative Livelihoods, June 2005), p. 85. Also see, for example, The Senlis Group, *Helmand at War: The Changing Nature of the Insurgency in Southern Afghanistan and Its Effects on the Future of the Country* (London: The Senlis Council, 2006).

[30] William Byrd and Christopher Ward, *Drugs and Development in Afghanistan* (Washington, D.C.: World Bank, 2004).

[31] Barnett R. Rubin, *Road to Ruin: Afghanistan's Booming Opium Industry* (New York: Center on International Cooperation, New York University, 2004), p. 10.

[32] RAND-MIPT Terrorism Incident Database. Following are the yearly figures on insurgent-initiated attacks in Afghanistan: 2002 (65 attacks); 2003 (148 attacks); 2004 (146 attacks); 2005 (207 attacks); 2006 (353 attacks). Following are the fatalities during the same period: 2002 (79 deaths); 2003 (133 deaths); 2004 (230 deaths); 2005 (288 deaths); 2006

incorporate insurgent-initiated attacks against Afghan civilians, international aid workers, and coalition forces. The U.S. military reported that the increase in violence was particularly acute between 2005 and 2006. During this period, the number of suicide attacks increased by more than 400 percent (from 27 to 139), remotely detonated bombings more than doubled (from 783 to 1,677), and armed attacks nearly tripled (from 1,558 to 4,542).[33] In 2007, insurgent-initiated violence rose another 27 percent from 2006 levels. Helmand Province witnessed among the highest levels of violence, with a 60-percent rise between 2006 and 2007.[34] The result was a lack of security for Afghans and foreigners, especially those living in the east and south. Road travel in many areas was dangerous, and crime was a major problem. Interfactional— or "green-on-green"—fighting continued among regional commanders, including those in the provinces of Herat, Nangarhar, Nuristan, Logar, Laghman, and Badghis. As one report by the National Directorate for Security concluded, Taliban cells in the south of Afghanistan developed good intelligence about individuals in villages and towns,

> Individuals who flirt with the government truly get frightened as the Afghan security forces are currently incapable of providing police and protection for each village . . . . When villagers and rural communities seek protection from police either it arrives late or arrives in a wrong way.[35]

Promoting disorder among the population is a key objective of most insurgents. Disrupting the economy and decreasing security helps

---

(755 deaths). A comparison of the RAND-MIPT data with U.S. and European government data shows that the RAND-MIPT data significantly understate the number of attacks and deaths, since most improvised explosive device (IED) and armed attacks were never reported in the press. Nevertheless, the trend in the RAND-MIPT data is consistent with U.S. and European government data.

[33] Pamela Constable, "Gates Visits Kabul, Cites Rise in Cross-Border Attacks," *Washington Post*, January 17, 2007, p. A10.

[34] These figures came from Admiral Michael Mullen, Chairman of the Joint Chiefs of Staff. See, for example, Ed Johnson, "Gates Wants NATO to Reorganize Afghanistan Mission," Bloomberg News, December 12, 2007.

[35] Saleh, *Strategy of Insurgents and Terrorists in Afghanistan*, p. 4.

produce discontent with the indigenous government and undermines its strength and legitimacy. Once insurgents establish a hold over the population, those who are hostile to the insurgents often become too fearful to oppose them. Some may be eliminated, providing an example to others. Some may escape abroad. Still others may be cowed into hiding their true feelings. By threatening the population, the insurgents give individuals a strong rationale to refuse or refrain from cooperating with the indigenous government and external actors.[36]

As learning organizations, insurgent groups were successful at continually adapting their tactics, techniques, and procedures to confront counterinsurgency efforts.[37] They conducted a wide variety of attacks against U.S., coalition, and Afghan security forces, as well as Afghan and international civilians. The insurgents relied heavily on asymmetric tactics, some of which were similar to those used by mujahideen forces against Soviet and Democratic Republic of Afghanistan army forces during the Soviet-Afghan war.[38] Insurgent tactics included yielding the population centers to U.S. and Afghan forces, operating from rural areas, distributing propaganda to the local population and opposition forces, threatening and intimidating the local population, and conducting armed attacks. As Taliban military officials argued, this is classic guerrilla warfare: "Our military tactic is to control a district center, kill the government soldiers there, and withdraw to our

---

[36] Galula, *Counterinsurgency Warfare*, pp. 11–12, 78–79.

[37] On terrorism and learning, see Brian A. Jackson, John C. Baker, Peter Chalk, Kim Cragin, John V. Parachini, and Horacio R. Trujillo, *Aptitude for Destruction,* Volume 1: *Organizational Learning in Terrorist Groups and Its Implications for Combating Terrorism* (Santa Monica, Calif.: RAND Corporation, MG-331-NIJ, 2005); Brian A. Jackson, John C. Baker, Peter Chalk, Kim Cragin, John V. Parachini, and Horacio R. Trujillo, *Aptitude for Destruction,* Volume 2: *Case Studies of Organizational Learning in Five Terrorist Groups* (Santa Monica, Calif., RAND Corporation, MG-332-NIJ, 2005).

[38] Grau, *The Bear Went Over the Mountain*; Grau, *Artillery and Counterinsurgency*; U.S. Army Training and Doctrine Command, *Operation Enduring Freedom: Tactics, Techniques, and Procedures* (Fort Leavenworth, Kan.: U.S. Army Training and Doctrine Command, December 2003).

mountainous strongholds, where it would be very difficult for the government to pursue us."[39]

Examples of armed attacks by the insurgency included ambushes and raids using small arms and grenades; shelling using 107-mm and 122-mm rockets and 60-, 82-, and 120-mm mortars; and improvised explosive devices (IEDs).[40] Most of their shelling and rocket fire was not accurate, though there is some evidence that insurgent forces considered harassment of enemy forces and populations as valuable. Insurgent groups, especially the Taliban, also succeeded in capturing government installations, villages, and district centers in the south, though usually for brief periods.

Taliban forces deployed in larger numbers over time, especially in such southern provinces as Helmand. In 2002, they operated in squad-size units. In 2005, they operated in company-sized units of up to 100 or more fighters. By 2008, they occasionally operated in battalion-sized units, though they deployed in smaller units as well.[41] This suggests that the Taliban were able to move around with more freedom in the south without being targeted by Afghan or coalition forces as time wore on. They also shifted from hard targets, such as U.S. forces, to soft targets, such as Afghan police and international personnel perceived to be supporting the Afghan government or coalition forces. Examples include Afghans organizing or otherwise involved in election work, NGO workers, ANP, ANA, and Afghan citizens believed to be cooperating with coalition forces or the Afghan government. Major spikes in insurgent-initiated violence were usually a function of spe-

---

[39] "Taliban Military Chief Threatens to Kill U.S. Captives."

[40] U.S. Marine Corps, *After Action Report on Operations in Afghanistan* (Camp Lejeune, N.C.: United States Marine Corps, August 2004); U.S. Army Training and Doctrine Command, *Operation Enduring Freedom.*

[41] LTG Karl Eikenberry, *Statement of Lt. Gen. Karl Eikenberry, Commander, Combined Forces Command—Afghanistan, Testimony Before the House Armed Services Committee,* June 28, 2006; GEN Barry R. McCaffrey (ret.), "Trip to Afghanistan and Pakistan," memorandum from General McCaffrey to COL Mike Meese and COL Cindy Jebb, United States Military Academy, June 2006, p. 4; U.S. Army Training and Doctrine Command, *Operation Enduring Freedom*; International Security Assistance Force, *Opposing Militant Forces: Elections Scenario* (Kabul: ISAF, 2005).

cific campaigns. Examples include the insurgent attempt to destabilize the October 2004 presidential elections and 2005 parliamentary elections in Afghanistan by targeting Afghan and international personnel involved in organizing, registering, and participating in the elections. Insurgents also conducted a major campaign tied to the U.S. handover of the counterinsurgency campaign to NATO in 2006. Attacks occurred throughout the country before and after the handover, though most were in the south and east around the provinces of Helmand, Paktia, Paktika, and Kandahar.[42]

Some of the most brutal executions conducted by the insurgents were of "collaborators" with the Afghan government or coalition forces.[43] These targets included the assassination of Islamic clerics critical of the Taliban, such as Mullah Abdullah Fayyaz, head of the Ulema Council of Kandahar.[44] As Figure 4.3 shows, primary targets included Afghan government officials, Afghan citizens, NGOs, educational institutions, and religious figures. Schools were increasingly targeted in such provinces as Helmand. As one Taliban night letter warned: "Teachers' salaries are financed by non-believers. Unless you stop getting wages from them, you will be counted among the American puppets."[45] This rationale also included targeting election candi-

---

[42] RAND-MIPT Incident Database; Michael Bhatia, Kevin Lanigan, and Philip Wilkinson, *Minimal Investments, Minimal Results* (Kabul: Afghanistan Research and Evaluation Unit, June 2004), pp. 1–8; Anthony Davis, "Afghan Security Deteriorates as Taliban Regroup," *Jane's Intelligence Review*, Vol. 15, No. 5 (May 2003), pp. 10–15.

[43] Al Qaeda in Afghanistan, *The Rule of Allah*, video, produced in 2006; "Taliban Execute Afghan Woman on Charges of Spying for U.S. Military," Afghan Islamic Press, August 10, 2005; "Afghan Taliban Report Execution of Two People on Charges of Spying for U.S.," Afghan Islamic Press, July 12, 2005.

[44] "Taliban Says Responsible for Pro-Karzai Cleric's Killing, Warns Others," *The News* (Islamabad), May 30, 2005; "Taliban Claim Responsibility for Killing Afghan Cleric," Kabul Tolu Television, May 29, 2005. Also see the killing of other clerics, such as Mawlawi Mohammad Khan, Mawlawi Mohammad Gol, and Mawlawi Nur Ahmad in "'Pro-Karzai' Cleric Killed by Bomb in Mosque in Khost Province," Pajhwok Afghan News, October 14, 2005; "Karzai Condemns Murder of Clerics," Pajhwok Afghan News, October 18, 2005.

[45] "Taliban Threatens Teachers, Students in Southern Afghan Province," Pajhwok Afghan News, January 3, 2006. Also see "Gunmen Set Fire to Schools in Ghazni, Kandahar Provinces," Pajhwok Afghan News, December 24, 2005.

Figure 4.3
Insurgent Targets, 2002–2006

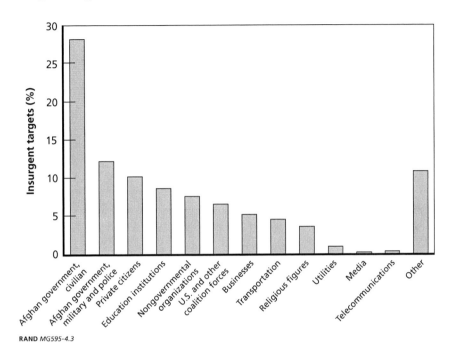

RAND MG595-4.3

dates and members of parliament, since "the elections are a part of the American program" and those who participate in the elections "are the enemies of Islam and the homeland."[46]

External Support

One of the most significant reasons for the insurgents' success in perpetrating a greater amount of violence was the support they received from two types of external actors: states and the international jihadist movement. In most insurgencies, the insurgent's ability to achieve sanctuary in neighboring countries presents a major challenge for indigenous governments. Opportunities for such sanctuary are often exploited

---

[46] Afghan Islamic Press, Interview with Mofti Latifollah Hakimi, August 30, 2005.

by insurgents. It is more difficult for counterinsurgent forces to target insurgents who have retreated to these sanctuaries, which allow the insurgents to regroup, resupply, and recruit new members. Statistical evidence shows that mountainous terrain can provide a particularly useful sanctuary for insurgent groups because it is difficult for indigenous and external counterinsurgent forces to navigate and easier for insurgents to hide in.[47] This presented a particular challenge in Afghanistan, since the border areas and sanctuary in Pakistan included some of the world's most rugged, mountainous terrain.[48]

Support from External States

Insurgent groups were successful at leveraging assistance from external states—especially in Pakistan. There are indications that support from Pakistan included two major components: assistance from some officials in the Pakistan government and the freedom to operate on Pakistani soil.

Officials in the Pakistan government had ideological and geostrategic motivations. As Pakistan's former dictator General Zia-ul-Haq once remarked to the head of the ISI, General Akhter Abdul Rehman, "the water [in Afghanistan] must boil at the right temperature."[49] Afghanistan has long been important to Pakistan policymakers because of its geographic location. Following the overthrow of the Taliban regime in 2001, officials in the Pakistan government were motivated to work with the Taliban for several reasons:

- to balance against India, especially in light of Delhi's close relationship with the Afghan government

---

[47] Fearon and Laitin, "Ethnicity, Insurgency, and Civil War;" Galula, *Counterinsurgency Warfare*, pp. 35–37.

[48] On the impact of mountainous terrain on insurgencies, see Hosmer, *The Army's Role in Counterinsurgency and Insurgency*, pp. 30–31; Daniel L. Byman, Peter Chalk, Bruce Hoffman, William Rosenau, and David Brannan, *Trends in Outside Support for Insurgent Movements* (Santa Monica, Calif.: RAND Corporation, MR-1405-OTI, 2001); Byman, *Deadly Connections*.

[49] Quoted in Praveen Swami, "Covert Contestation," *Frontline*, Vol. 22, No. 19 (September 2005).

- to hedge against a U.S. and NATO withdrawal, ensuring that if Western troops departed from Afghanistan, Pakistan would retain a proxy force in Afghanistan
- to preempt a movement among Pakistan's Pashtun population toward closer relations with Afghanistan should Afghanistan became more secure and prosperous.

The ISI provided assistance to the Taliban in the 1990s and early 2000s to ensure that it had an ally in Kabul. The motivation of those in the ISI assisting the Taliban in the mid-2000s was similar: to increase the likelihood over the long term that the government of Afghanistan (or at least those controlling the areas of Afghanistan near Pakistan) was controlled by allies.

Balancing against India appeared to be a particularly strong impetus for Pakistan's support of the insurgents. Pakistan and India have long been involved in a balance-of-power struggle in South Asia. Both lay claim to the Kashmir region and fought at least three wars over Kashmir since 1947. Since September 11, 2001, India provided several hundred million dollars in financial assistance to Afghanistan, including funds to assist Afghan political candidates during the 2004 presidential elections and 2005 parliamentary elections. India also helped fund construction of the new Afghan parliament building and provided financial assistance to elected legislators.[50] India's road construction near the Pakistan border was a significant point of contention between India and Pakistan. These projects were run by India's state-owned Border Roads Organisation, whose publicly acknowledged mission was to help "the [Indian] armed forces meet their strategic needs by committed, dedicated and cost-effective development and sustenance of the infrastructure."[51] Finally, India established several consulates in such Afghan cities as Jalalabad, Kandahar, and Herat. Pakistan

---

[50] David C. Mulford, *Afghanistan Has Made a Remarkable Transition* (New Delhi: U.S. Department of State, February 2006); Amin Tarzi, "Afghanistan: Kabul's India Ties Worry Pakistan," Radio Free Europe/Radio Liberty, April 16, 2006. Mulford is U.S. Ambassador to India.

[51] Border Roads Organisation, *Vision, Mission, Role* (Delhi: Border Roads Organisation, 2006).

accused India of using these consulates as a base for "terrorist activities" conducted inside Pakistan, such as fomenting unrest in the province of Baluchistan. The Indian-Afghan axis left Pakistan isolated among its South Asian neighbors. Before the September 11, 2001, terrorist attacks in the United States, Pakistan had a close relationship with the Taliban government in Afghanistan, which it had nurtured since the 1990s. Half a decade later, Pakistan was surrounded by hostile states. Consequently, for some Pakistani officials, assisting insurgents in Afghanistan was a way to balance against Indian influence in Afghanistan, maximize Pakistan's influence in the border regions, and prevent the Pashtuns on both sides of the border from developing a unified front and pushing for integration into Afghanistan.

Some active and former Pakistan government officials from organizations such as the ISI and Frontier Corps provided logistical support to the Taliban and helped secure medical care for wounded insurgents in cities such as Quetta. They also helped train Taliban and other insurgents destined for Afghanistan and Kashmir in Quetta, Mansehra, Shamshattu, Parachinar, and other areas within Pakistan. To minimize its visibility, these individuals appeared to supply indirect assistance—including financial assistance—to Taliban training camps. NATO officials uncovered several instances in which ISI operatives provided intelligence to Taliban insurgents at the tactical, operational, and strategic levels. This included tipping off Taliban forces about the location and movement of Afghan and coalition forces, which undermined several U.S. and NATO anti-Taliban military operations.[52]

In addition, General Hamid Gul and Colonel Sultan Amir Imam, pro-Taliban and pro–al Qaeda Pakistani leaders, gave widely reported speeches at government and military institutions in Pakistan calling for jihad against the United States and the Afghan government.[53] In

---

[52] There is an extensive unclassified literature discussing support for insurgents by some individuals within the Pakistan government. See, for example, Carlotta Gall and David Rohde, "Militants Escape Control of Pakistan, Officials Say," *New York Times*, January 15, 2007, p. A1; Rubin, "Saving Afghanistan;" Arnaud de Borchgrave, "Talibanization of Pakistan," *Washington Times*, April 7, 2007, p. A11; and Seth G. Jones, "Pakistan's Dangerous Game," *Survival*, Vol. 49, No. 1, Spring 2007, pp. 15–32.

[53] Rubin, *Afghanistan and the International Community*, p. 24.

sum, individuals within the ISI and other Pakistan government agencies provided several types of assistance:

- ensuring that wounded Taliban and other insurgents received medical aid
- training insurgents at camps in Pakistan
- providing intelligence
- providing financial assistance
- assisting with logistics in crossing the border.

This assistance is consistent with the Pakistan government's past behavior, especially the ISI. Throughout the 1990s, Pakistan's military and intelligence service provided arms, ammunition, supplies, financial aid, and training to the Taliban and Gulbuddin Hekmatyar. Pakistan also helped recruit fighters for the Taliban, sometimes working with domestic religious associations.[54]

In addition, insurgent groups had substantial freedom to operate in Pakistan. The Taliban and other insurgent groups shipped arms, ammunition, and supplies into Afghanistan from Pakistan. Many suicide bombers came from Afghan refugee camps located in Pakistan. IED components were often smuggled across the Afghanistan-Pakistan border and assembled at safe houses in and around such provinces as Kandahar. The Taliban used roads such as Highway 4 in Kandahar Province to transport fighters and supplies between Afghanistan and Pakistan.[55]

Pakistan's government failed several times to negotiate effective peace deals with militants in such tribal regions as North and South Waziristan.[56] These deals called on tribesmen to expel foreign militants and end cross-border attacks into Afghanistan. In return, Pakistan's

---

[54] Byman, *Deadly Connections*, pp. 194–198; Coll, *Ghost Wars*; Roy, *Islam and Resistance in Afghanistan*; Rashid, *Taliban*; Maley, *Fundamentalism Reborn?*

[55] Author interviews with NATO officials in Afghanistan, January 2007.

[56] See, for example, "Peace Pact: North Waziristan," September 5, 2006. This agreement was negotiated by a "political agent from North Waziristan representing Governor N.W.F.P. Federal Government," and "tribal representatives from North Waziristan, Local Mujahideen N.W.F.P, Atmanzai Tribe"; N.W.F.P. is Pakistan's North West Frontier Province.

military promised to end major operations in the area and pull most of its soldiers back to military camps. The logic of these deals seems intuitive: In areas where tribes exert political, military, and economic power, the most effective long-term solution was to create incentives for tribal leaders to police their areas. After all, these tribal areas had been ruled indigenously for hundreds of years. And tribes often regard outside forces, including Pakistan's military, as unwelcome foreigners.

But there were several problems with this strategy. First, it rested on a false assumption, since it presumed that tribes actually controlled these areas. A closer look at the tribal areas indicated that insurgents and terrorists like the Taliban increasingly exerted control. In many cases, they usurped the power of tribes. Expecting tribes to police areas they did not even control was wishful thinking. Second, the tribal deals failed to curb cross-border activity and undermine the power of the Taliban and other militant groups there. NATO officials I interviewed argued that insurgents crossed the border in greater numbers.[57] As former Pakistan foreign minister Najmuddin Shaikh acknowledged in Pakistan's *Dawn* newspaper, "There is no doubt that the Waziristan agreement has led to increased Taliban influence."[58] Third, there was no enforcement mechanism were the tribal deals to fail. Why should tribes cooperate, assuming they could, if there was no penalty for defection? Pakistan's military expressed a deep unwillingness to enter the tribal areas again.

Indeed, Afghan insurgents used Pakistan as a staging area for offensive operations. Taliban insurgents that operated in the southern Afghan provinces of Kandahar, Oruzgan, Helmand, and Zabol had significant support networks in such Pakistani provinces as Baluchistan and the Federally Administered Tribal Areas, including in Waziristan. Due to common ethnicity, they received political support from some of Pakistan's Pashtun tribes.[59] The Taliban conduct much

---

[57] Author interview with senior NATO officials, Afghanistan, 2006 and 2007.

[58] Najmuddin A. Shaikh, "Worsening Ties with Kabul," *Dawn* (Pakistan), December 13, 2006.

[59] Author interview with Afghan Foreign Minister Rangin Dadfar Spanta, Washington, D.C., July 2006; LTG David W. Barno, *Afghanistan: The Security Outlook* (Washington,

of their financing and recruiting operations on the Pakistani side of the border.[60] There is also significant evidence that the Taliban leadership had a support base—commonly referred to as the Quetta Shura—in Quetta, Pakistan.[61] As Zalmay Khalilzad, former U.S. Ambassador to Afghanistan, noted,

> Mullah Omar and other Taliban leaders are in Pakistan. [Mullah Akhtar] Usmani, one of the Taliban leaders, spoke to Pakistan's Geo TV at a time when the Pakistani intelligence services claimed that they did not know where [the Taliban leaders] were. If a TV company could find him, how is it that the intelligence service of a country which has nuclear bombs and a lot of security and military forces cannot find them?"[62]

In addition, Ali Jalali, former Afghan Interior Minister, argued,

> The Taliban have training camps, staging areas, recruiting centers (*madrassas*), and safe havens in Pakistan. The operations of a 70,000-strong Pakistani military force, deployed in the border region, mostly in the Waziristan tribal areas, have been effective against al Qaeda and non-Pakistani militants, but they have not done much toward containing the Taliban.[63]

D.C.: Center for Strategic and International Studies, May 14, 2004); Center for Army Lessons Learned, *Ranger Observations from OEF and OIF: Tactics, Techniques, and Procedures* (Fort Leavenworth, Kan.: Center for Army Lessons Learned, February 2005), p. 21; David L. Buffaloe, *Conventional Forces in Low-Intensity Conflict: The 82nd Airborne in Firebase Shkin*, Landpower Essay 04-2 (Arlington, Va.: Association of the United States Army), pp. 16–17.

[60] Ahmed Rashid, "Who's Winning the War on Terror?" *YaleGlobal*, September 5, 2003.

[61] Author interview with officials from several Western government agencies, 2005 and 2006.

[62] Zalmay Khalilzad, "Outgoing U.S. Envoy Enthusiastic About Afghanistan's Future," interview on Sherberghan Jowzjan Aina Television, June 18, 2005. Ambassador Khalilzad's comments were supported by President Karzai's office ("Afghan Spokesman Calls on Pakistan to Curb Taliban Activities," Kabul Tolu Television, June 21, 2005).

[63] Jalali, "The Future of Afghanistan," p. 8.

The Pakistani military conducted combat operations against foreign fighters—especially Central Asians and Arabs—in the Federally Administered Tribal Areas.[64] But Pakistan was reluctant to conduct operations in Baluchistan against Taliban insurgents or their support network.[65] As one Pakistani journalist argued, the Pakistan government "plunges into action when they know they can lay their hands on a foreign militant but they are still reluctant to proceed against the Taliban."[66] Part of the reason may be that the Pakistan government was preoccupied with other security concerns in such provinces as Baluchistan, where it was fighting a counterinsurgency campaign against Baluch tribes.[67]

In addition to Pakistan, Iran has historically been active in Afghanistan. Iranian policymakers have long been interested in securing strategic depth and influence in Afghanistan. Following the overthrow of the Taliban regime in 2001, the Iranian government funded reconstruction projects in Afghanistan (including road construction projects) and provided aid to some warlords. This behavior was consistent with the activities of such regional powers as Pakistan, India, and Russia. Indeed, the Iranian strategy in Afghanistan after the overthrow of the Taliban is perhaps best characterized as a "hedging strategy."[68]

---

[64] "Pakistan Strikes Suspected al Qaeda Camp," Associated Press, March 1, 2006.

[65] There have been some notable exceptions, such as the Pakistani government's capture of Taliban spokesman Abdol Latifollah Hakimi in October 2005. The Pakistani government has also closed a few businesses owned by Taliban-linked traders. In January 2006, for example, the Pakistani government closed Haji Abdul Bari's Special Company in Peshawar and the Maria Food Company in Islamabad, and froze $5 million in their accounts, because the owners helped finance Taliban fighters.

[66] Intikhab Amir, "Waziristan: No Man's Land," *The Herald* (Pakistan), April 2006, p. 78.

[67] Shahzada Zulfiqar, "Endless War," *The Herald* (Pakistan), April 2006, pp. 33–36; Frederic Grare, *Pakistan: The Resurgence of Baluch Nationalism* (Washington, D.C.: Carnegie Endowment for International Peace, January 2006).

[68] On hedging, see Robert J. Art, "Europe Hedges Its Security Bets," in *Balance of Power: Theory and Practice in the 21st Century*, ed. T. V. Paul, James Wirtz, and Michel Fortmann (Palo Alto, Calif.: Stanford University Press, 2004), pp. 179–213; Randall L. Schweller, "Managing the Rise of Great Powers: History and Theory," in Alastair Iain Johnston and Robert S. Ross, eds., *Engaging China: The Management of Emerging Power* (New York: Routledge, 1999), pp. 1–32.

The Iranian government preferred a close relationship with the Afghan government, which it enjoyed with key Afghan policymakers. Iran and Afghanistan cooperated in trying to crack down on drugs passing over their shared border. They also participated in joint trade, energy, investment, cultural, and scientific projects.[69]

However, Iran also saw its involvement in Afghanistan as a hedge against a possible U.S. or Israeli strike against Iranian nuclear facilities. This meant that Iran was prepared to undermine U.S. efforts in Afghanistan in the event of further deterioration of U.S.-Iranian relations. There was some evidence that individuals from the Iranian government, including from the Iranian Revolutionary Guard Corps–Quds Force, provided some arms, money, and training to Taliban commanders and other insurgents. Examples included explosively formed penetrators, anti-tank mines, mortars, and small arms.[70]

However, there were limits to Iran's willingness to support the Taliban and other insurgent groups. Iran historically had poor relations with the Taliban. As one Iranian diplomat noted in 1997, for example, Iran joined "Russia and the anti-Taliban alliance against Pakistan, Saudi Arabia and the Taliban" because it was concerned about the rise of Sunni extremist groups like the Taliban.[71]

Jihadist Support
Another source of support for the insurgency in Afghanistan was the international jihadist network, which enabled the Taliban and other groups to sustain their operations and helped them become more lethal in attacks against Afghans and coalition forces. This support came from a variety of different jihadist sources. One source was organiza-

---

[69] Author interview with Afghan Foreign Minister Rangin Dadfar Spanta, Washington, D.C., July 2006; author interview with Zalmai Rassoul, Afghanistan National Security Advisor, Kabul, Afghanistan, November 2005.

[70] On Iranian involvement in Afghanistan, see McConnell, 2008; John Ward Anderson, "Arms Seized in Afghanistan Sent from Iran, NATO Says," *Washington Post*, September 21, 2007, p. A12; Robin Wright "Iranian Destined for Taliban Seized in Afghanistan," *Washington Post*, September 16, 2007, p. A19; and Jason Motlagh, "Weapons Convoy from Iran Reported." *Washington Times*, June 22, 2007, p. A13.

[71] Quoted in Rashid, *Taliban*, p. 177.

tions such as the international al Qaeda network. Afghan insurgent groups also received assistance from the collection of *zakat* (the Islamic concept of tithing and alms) at mosques in Pakistan, Afghanistan, and the broader Muslim world. Finally, much of the jihadist funding came from wealthy Muslims abroad, especially from such Gulf states as the United Arab Emirates, Saudi Arabia, and Qatar. For example, al Qaeda personnel regularly met with wealthy Arab businessmen during the Tabligh Jamaat annual meeting in Raiwind, Pakistan, which attracted one of the largest concentrations of Muslims after the hajj.[72]

The Taliban and other insurgent groups established a major support base through their cooperation with Jamiat-e-Ulema Islam. This Pakistani political party had its roots in the Deobandi movement and had a following largely confined to the Pashtun border belt of the North West Frontier Province and Baluchistan (although it also has support in several of Pakistan's urban centers). The Deobandi movement developed in British-ruled India during the mid-1800s. It was an offshoot of the Sunni Hanafi legal school and took its name from the Indian Himalayan town of Deoband, the location of an influential religious school. The Deobandi movement aimed to reform and unify Muslims, preached strict adherence to the *Sunnah* (the way or deeds of the Prophet Muhammad), and emphasized the importance of *shari'a* law. Jamiat-e-Ulema Islam was split into two factions, led by Maulana Fazal ur-Rehman and Samiul Haq (a fervent supporter of Osama bin Laden). The party ran an extensive network of madrassas that trained most of the leadership and much of the early rank and file of the Taliban. Party links with the Taliban remained close, despite President Musharraf's talk of reforming the madrassas. Indeed, Afghan insurgents long targeted recruits at madrassas and Afghan refugee camps in Pakistan.

Al Qaeda played a critical role in the insurgency as a force multiplier, assisting insurgent groups such as the Taliban at the tactical, operational, and strategic levels. Groups such as the Taliban used sup-

---

[72] Alex Alexiev, "Tablighi Jamaat: Jihad's Stealthy Legions," *Middle East Quarterly*, Vol. XII, No. 1 (Winter 2005). On zakat and jihad, also see Marc Sageman, *Understanding Terror Networks* (Philadelphia: University of Pennsylvania Press, 2004).

port and training from jihadists to construct increasingly sophisticated IEDs, including IEDs with remote-control detonators. For example, there were a handful of al Qaeda–run training facilities and IED assembly facilities in such places as North and South Waziristan. They ranged from small facilities hidden in compounds to much larger "IED factories," which doubled as training centers and labs where recruits experimented with IED technology. These facilities were located in such remote places as the Bush mountains, Khamran mountains, and Shakai valley. Al Qaeda received operational and financial support from local clerics and Taliban commanders in Waziristan. They recruited young Pashtuns from the local madrassas and financed their activities through "religious racket"—forced religious contribution, often accompanied with death threats. Some of this IED expertise came from Iraqi groups, which provided information to Afghan groups on making and using various kinds of remote-controlled devices and timers. Indeed, there is evidence of cooperation between insurgents in Iraq and Afghanistan. Islamic militants in Iraq provided information through the Internet and face-to-face visits on tactics to Taliban, Hezb-i-Islami, and foreign fighters from eastern and southern Afghanistan and Pakistan's tribal areas. In addition, there is some evidence that a small number of Pakistani and Afghan militants received military training in Iraq; Iraqi fighters met with Afghan and Pakistani extremists in Pakistan; and militants in Afghanistan increasingly used homemade bombs, suicide attacks, and other tactics honed in Iraq.[73]

The "TV bomb" is one example of an IED introduced to Afghan insurgents by Iraqi groups. This shaped-charge mechanism can be hidden under brush or debris on a roadside and set off by remote control from a distance of 300 yards or more. There is also some evidence that individuals such as Hamza Sangari, a Taliban commander from Khowst Province, received information from Iraqi groups that improved the Taliban's ability to make armor-penetrating weapons by disassembling rockets and rocket-propelled grenade rounds, remov-

---

[73] Interview with Mullah Dadullah, Al Jazeera TV, July 2005. Also see such press accounts as Sami Yousafzai and Ron Moreau, "Unholy Allies," *Newsweek*, September 26, 2005, pp. 40–42.

ing the explosives and propellants, and repacking them with high-velocity "shaped" charges.[74] Afghan groups occasionally adopted some of the more brutal tactics, such as beheadings, used by Iraqi groups. In December 2005, for example, insurgents posted a video to al Qaeda–linked Web sites showing the decapitation of an Afghan hostage—the first time a video of the beheading of an Afghan hostage was shown.[75] The Taliban also acquired new commercial communication gear and field equipment from Iraqi groups and received good tactical, camouflage, and marksmanship training from them as well.

In addition, Afghan insurgents increasingly adopted suicide tactics, especially in major cities such as Kandahar and Kabul.[76] The number of suicide attacks increased steadily: one in 2002, two in 2003, six in 2004, and twenty-seven in 2005. There were 139 suicide terrorist attacks in Afghanistan in 2006 and 140 in 2007.[77] The use of suicide attacks was encouraged by al Qaeda leaders in Pakistan, such as Ayman al-Zawahiri, who argued for the "need to concentrate on the method of martyrdom operations as the most successful way of inflicting damage against the opponent and the least costly to the Mujahedin in terms of casualties."[78] Suicide bombers included Afghans, Pakistanis, and some foreigners.[79] Most suicide bombers through 2007 came from Afghan refugee camps in Pakistan. They frequently attended Pakistani

---

[74] Author interview with Afghan government officials, Kabul, Afghanistan, August 2006.

[75] In what appeared to be a forced confession, Saeed Allah Khan stated, "I worked as a spy for the Americans along with four other people. The group received $45,000 and my share is $7,000" (Hekmat Karzai, *Afghanistan and the Globalisation of Terrorist Tactics,* Singapore: Institute of Defence and Strategic Studies, January 2006).

[76] On the rationale for suicide bombers, see Interview with Mullah Dadullah, Al Jazeera TV, February 2006.

[77] The data on suicide attacks came from the RAND-MIPT Terrorism Incident Database, the U.S. Department of Defense, the Centre for Conflict and Peace Studies in Kabul, and Jason Straziuso, "U.S. Casualties in Afghanistan Hit Record," *Navy Times,* January 2, 2008.

[78] Ayman al-Zawahiri, *Knights Under the Prophets Banner* (n.p., December 2001).

[79] In its public rhetoric, the Taliban has tended to identify the suicide bombers as Afghans, since it suggests that there is a significant indigenous component of the insurgency.

madrassas, where they were radicalized and immersed in extremist ideologies.[80]

Several factors can be attributed to the rise in suicide attacks.[81] First, the Taliban successfully tapped into the expertise and training of the broader jihadist community, especially al Qaeda. Jihadists imparted knowledge on suicide tactics to Afghan groups through the Internet and in face-to-face visits. With al Qaeda's assistance, these militants helped supply a steady stream of suicide bombers. Second, al Qaeda and the Taliban concluded that suicide bombing was more effective than other tactics in killing Afghan and coalition forces. This was a direct result of the success of such groups as Hamas in the Palestinian territories, Hezbollah in Lebanon, the Tamil Tigers in Sri Lanka, and Iraqi groups.[82] Suicide attacks allowed insurgents to achieve maximum impact with minimal resources. Data show that when insurgents fight U.S. and coalition forces directly in Afghanistan, there is only a 5 percent probability of inflicting casualties. With suicide attacks, the chance of killing people and instilling fear increased several fold.[83] Third, al Qaeda and the Taliban believed that suicide attacks increased the level of insecurity among the Afghan population. This caused some Afghans to question the government's ability to protect them and further destabilized

---

[80] Author interviews with NATO officials in Kandahar, Afghanistan, January and September 2007; United Nations Assistance Mission to Afghanistan, *Suicide Attacks in Afghanistan: 2001–2007* (Kabul: United Nations Assistance Mission to Afghanistan, 2007).

[81] Hekmat Karzai, *Afghanistan and the Logic of Suicide Terrorism* (Singapore: Institute of Defence and Strategic Studies, March 2006); "Taliban Claim Responsibility for Suicide Bomb Attack in Afghan Kandahar Province," Peshawar Afghan Islamic Press, October 9, 2005; "Pajhwok News Describes Video of Afghan Beheading by 'Masked Arabs,' Taliban," Pajhwok Afghan News, October 9, 2005; "Canadian Soldier Dies in Suicide Attack in Kandahar," Afghan Islamic Press, March 3, 2006; "Taliban Claim Attack on Police in Jalalabad, Nangarhar Province," Kabul National TV, January 7, 2006.

[82] See, for example, Robert Pape, *Dying to Win: The Strategic Logic of Suicide Terrorism* (New York: Random House, 2005); Mia Bloom, *Dying to Kill: The Allure of Suicide Terror* (New York: Columbia University Press, 2005); Christoph Reuter, *My Life Is a Weapon: A Modern History of Suicide Bombing* (Princeton, N.J.: Princeton University Press, 2004); Hoffman, *Inside Terrorism*.

[83] Hekmat Karzai and Seth G. Jones, "How to Curb Rising Suicide Terrorism in Afghanistan," *Christian Science Monitor*, July 18, 2006, p. 9.

the authority of local government institutions. Fourth, suicide attacks provided renewed visibility for the Taliban and al Qaeda, which previous guerrilla attacks did not generate. Because of their lethality and high-profile nature, every suicide attack was reported in the national and international media.

In sum, the international jihadist network provided significant support to Afghan insurgent groups. Al Qaeda effectively spread its extremist global ideology in Afghanistan and Pakistan. It played a critical role in providing encouragement and impetus for the utilization of suicide attacks and sophisticated IEDs. Al Qaeda also paid up to several thousand dollars to the families of suicide bombers who perished in operations in Afghanistan. In addition, some Taliban units included al Qaeda members or other Arab fighters, who brought with them tactics employed in such places as Iraq and Chechnya.[84]

Conclusion

External support from state and nonstate actors was critical to the return of the Taliban and the rise of the insurgency in Afghanistan beginning in 2002. Support came from individuals in the Pakistan government and the international jihadist movement, including al Qaeda. Insurgent groups with significant external support and well-established sanctuaries in neighboring states have frequently been successful. This is a recurring lesson across insurgencies since at least 1945. Chapter Seven will explore in more depth U.S. capabilities that might be useful to counter external support networks.

---

[84] Author interview with U.S. government officials, Kabul, Afghanistan, December 2005.

CHAPTER FIVE

Afghan Government and Security Forces

This chapter examines lessons from the conduct of the Afghan government and security forces. It argues that two critical variables related to the indigenous government impact the success or failure of counterinsurgency operations: the quality of indigenous forces and governance capacity. Key Afghan forces include the ANP, ANA, and a range of allied militia forces such as the Afghan National Auxiliary Police. Building on the argument made in the previous chapter (that insurgents were increasingly able to conduct violence in the south and east of Afghanistan due in large part to external support and sanctuary), this chapter contends that Afghan forces were a mixed bag as far as their ability to establish security. The ANA was relatively competent in combat operations, while the police were largely incompetent and corrupt. This chapter also contends that the weakness of Afghanistan's central government was problematic. Corruption within the government was detrimental to the counterinsurgency campaign and decreased popular support for the Afghan government.

Lessons from the Afghan government and its security forces can be grouped into two broad categories: the capability of indigenous forces (including Afghanistan's intelligence agency) and the governance capacity of national, provincial, and district government institutions.

Capability of Indigenous Forces

The competence of Afghanistan's indigenous security forces is difficult to judge for two reasons. First, many of the most useful metrics are qualitative rather than quantitative and difficult to measure accurately. Examples include the performance of security forces in conducting cordon-and-search operations, patrols, border security, riot control, intelligence collection, and combat operations.[1] Second, little data have been systematically collected for several decades in Afghanistan. For example, there are no reliable statistics on homicide rates, which may provide some indication of the competence of police forces. We therefore must resort to largely qualitative judgments.

Afghan National Police

The evidence suggests that ANP was the least competent of the Afghanistan government forces. The first team of German police trainers arrived in Kabul in March 2002 to train police instructors. This training was critical, since Afghan police had not received any formal training for at least two decades.[2] The Germans focused on training inspectors and lieutenants at the police academy in Kabul. Officers went through a three-year training course and took classes on human rights, tactical operations, narcotics investigations, traffic, criminal investigations, computer skills, and Islamic law.[3] By 2003, however, U.S. officials from the Department of State, Department of Defense, and the White House became increasingly unhappy with the German approach. Many argued that it was far too slow, trained too few police officers, and was seriously underfunded.[4] Consequently, the

---

[1] See, for example, Combined Forces Command—Afghanistan, *Afghan National Security Forces Operational Primacy Process* (Kabul, Afghanistan: Combined Forces Command—Afghanistan, 2006).

[2] Government of Germany, Federal Foreign Office and Federal Ministry of Interior, *Assistance in Rebuilding the Police Force in Afghanistan*, p. 6; Asian Development Bank and World Bank, *Afghanistan: Preliminary Needs Assessment for Recovery and Reconstruction* (Kabul: Asian Development Bank and World Bank, January 2002), p. 7.

[3] Author interview with Jochen Rieso, Training Branch, German Project for Support of the Police in Afghanistan, Kabul, Afghanistan, June 27, 2004.

[4] As one high-level U.S. official noted, "When it became clear that they were not going to provide training to lower-level police officers, and were moving too slowly with too few resources, we decided to intervene to prevent the program from failing" (author interview with senior U.S. official, White House, September 2004). This view was corroborated by

United States concentrated on training lower-level recruits at a central training center in Kabul, as well as at regional training centers in outlying cities, such as Kandahar, Mazar-e Sharif, Gardez, and Jalalabad. The Department of State's Bureau of International Narcotics and Law Enforcement contracted DynCorp to train the police and to help build training facilities.[5] Beginning in 2005, the U.S. military took over the lead U.S. police training role after rising concerns in the U.S. Department of Defense about the effectiveness of the police program. Key problems included the failures to conduct follow-on mentoring of Afghan police, to provide significant institutional reform in the Ministry of Interior, and to curb deep-seated corruption in the police and Ministry of Interior.[6] The U.S. military provided training, equipment, and other assistance to ANP and internal security forces through the Combined Security Transition Command–Afghanistan.

The available evidence suggests that the ANP was corrupt and often unable to perform basic patrolling, conduct counterinsurgency operations, protect reconstruction projects, prevent border incursions, or conduct counternarcotics operations. Multiple interviews with U.S. and other NATO officials involved in police training from 2004 through 2008 indicate that corruption was pervasive in the ANP. Police regularly took bribes to allow drugs and other licit and illicit goods to pass along routes they controlled. Police chiefs were frequently involved in "skimming" money they received to pay their police officers.[7] Indeed, corruption appeared to be more pervasive in the police than in other security forces.[8]

---

multiple interviews by the author with U.S. officials in Washington and Afghanistan in 2004 and 2005.

[5] Author interview with employees of DynCorp, Kabul and Gardez, June 2004 and November 2005.

[6] Author interview with members of the Office of Security Cooperation–Afghanistan, Kabul, December 2005.

[7] Author interviews with U.S. and German police officials, Afghanistan, 2004, 2005, 2006, and 2007.

[8] Afghanistan Ministry of Interior, *Afghan National Police Program* (Kabul: Ministry of Interior, 2005); Barnett Rubin, *Afghanistan's Uncertain Transition from Turmoil to Normalcy* (New York: Council on Foreign Relations, 2006); U.S. Government Accountability Office,

ANP competence was also low. As a 2006 German assessment of the border police concluded, "Neither the Afghan border police nor the customs authorities are currently in a position to meet the challenges presented by this long border."[9] A report by the Offices of Inspector General of the U.S. Departments of State and Defense concluded that the "ANP's readiness level to carry out its internal security and conventional police responsibilities is far from adequate. The obstacles to establish a fully professional ANP are formidable." It found that key obstacles included "no effective field training officer (FTO) program, illiterate recruits, a history of low pay and pervasive corruption, and an insecure environment."[10]

ANP forces had a difficult time even against criminal organizations. In February 2006, for example, ANP forces were attacked, captured, and disarmed by a drug cartel in Balkh Province after an armed clash.[11] As Ali Jalali, former Afghan Minister of Interior, argued, "[B]ecause of the late start in comprehensive police development, the ANP continues to be ill-trained, poorly paid, under-equipped, and inadequately armed."[12] The ANP was vital to establishing order in urban and rural areas. But it was badly equipped, corrupt, poorly trained, and lacked any semblance of a national police infrastructure. There was little oversight at the provincial or district levels. The Afghan police lacked uniforms, armored vehicles, weapons, ammunition, police stations, police jails, national command and control, and investigative

Afghanistan Security: Efforts to Establish Army and Police Have Made Progress, but Future Plans Need to Be Better Defined (Washington, D.C.: GAO, 2005).

[9] Government of Germany, *Doha II Conference on Border Management in Afghanistan: A Regional Approach* (Berlin: Government of Germany, 2006). Also see U.S. Department of State, *Border Management Initiative: Information Brief* (Kabul: Afghanistan Reconstruction Group, U.S. Department of State, 2005).

[10] Offices of Inspector General of the Departments of State and Defense, *Interagency Assessment of Afghanistan Police Training and Readiness* (Washington, D.C.: Offices of Inspector General of the Departments of State and Defense, 2006), p. 1.

[11] Afghan Non-Governmental Organization Security Office, *Security Incident—Armed Clash: ANP Was Disarmed* (Kabul: Afghan Non-Governmental Organization Security Office, March 2006).

[12] Jalali, "The Future of Afghanistan," p. 10.

training.[13] These deficiencies impacted not only the counterinsurgency campaign, but also security more broadly.

Several factors contributed to the poor competence of the ANP. International training was not as good as it was for the ANA. The police were not an international priority in the early stages of the insurgency and received significantly less money and attention than the ANA. Interviews with U.S. and German officials involved in police training suggest that there were significant challenges with DynCorp, the U.S.-based company hired by the U.S. Department of State to implement police training at regional training centers across Afghanistan. One recurring criticism was wide variation in the quality of DynCorp police trainers. Some had significant international police training experience and were competent in dealing with police in a tribal society in the middle of an insurgency. But many other DynCorp trainers had little experience or competence.[14] The use of private contractors for police training or other tasks during counterinsurgency and nation-building operations has long been controversial.[15] At the very least, the U.S. government should conduct a thorough assessment of the performance and effectiveness of contractors in police training.

In addition, the lack of follow-on mentoring meant that police were given a few weeks of training and sent back to their villages with no oversight and assistance. Most had never received formal police training. ANA soldiers benefited from embedded international trainers when they deployed into the field, and the ANA almost always deployed with U.S. and other coalition military forces. But the ANP lacked a comprehensive mentoring program. U.S. officials argued even in 2008 that security concerns in the south and east precluded the deployment of mentoring teams to rural areas. This meant that in areas where a competent ANP was essential because of the growing presence

---

[13] McCaffrey, "Trip to Afghanistan and Pakistan."

[14] Author interviews with U.S. and German police officials, Afghanistan, 2004, 2005, 2006, 2007, and 2008.

[15] On the role of civilian contractors for security sector reform, see, for example, P. W. Singer, *Corporate Warriors: The Rise of the Privatized Military Industry* (Ithaca, N.Y.: Cornell University Press, 2003).

of the Taliban, there was virtually no mentoring. The lack of mentoring made police more susceptible to local warlords. Indeed, ANP forces were frequently more loyal to local warlords, tribal leaders, and even criminal networks than to the central government. This reflects the weakness of the central government and the strength of nonstate actors in much of the country.

In counterinsurgency operations, the police need to be involved in the community at all levels, such as monitoring border posts and patrolling cities, villages, and highways. Building the police in counterinsurgencies should be a higher priority than the creation of the army because the police are the primary arm of the government in towns and villages across the country. Unfortunately, this was not the case in Afghanistan. The ANP needed to be at the forefront of fighting insurgents, illegal border incursions, illicit drug trade, warlords, and organized crime in Afghanistan, but they were overwhelmed. Unfortunately, the situation in Afghanistan is not unique—building a competent and legitimate police force has been a major challenge in numerous counterinsurgency and stability operations.[16]

Afghan National Army

The available data suggest that the competence of ANA forces has improved since 2002, when training began. The United States was the lead nation for building the ANA, although French, British, Turkish, and other instructors from coalition countries were also involved.[17]

---

[16] On policing during counterinsurgency and stability operations, see Robert B. Oakley, Michael J. Dziedzic, and Eliot M. Goldberg, eds., *Policing the New World Disorder: Peace Operations and Public Security* (Washington, D.C.: National Defense University Press, 1998); Seth G. Jones, Jeremy M. Wilson, Andrew Rathmell, and K. Jack Riley, *Establishing Law and Order After Conflict* (Santa Monica, Calif.: RAND Corporation, MG-374-RC, 2005); Robert M. Perito, *Where Is the Lone Ranger When We Need Him? America's Search for a Post-conflict Stability Force* (Washington, D.C.: United States Institute of Peace, 2004); David H. Bayley, *Democratizing the Police Abroad: What to Do and How to Do It* (National Institute of Justice, June 2001).

[17] The author visited the Office of Military Cooperation–Afghanistan in 2004 and the Office of Security Cooperation–Afghanistan in 2005, as well as regional training centers, to assess the U.S. and coalition efforts to rebuild the ANA and ANP. On training the ANA, also see Anja Manuel and P. W. Singer, "A New Model Afghan Army," *Foreign Affairs*, Vol.

Training commenced in May 2002, when the ANA's first regular army battalion began 10 weeks of infantry and combat training at the Kabul Military Training Center. U.S. special operations forces assigned to the U.S. Army's 1st Battalion, 3rd Special Forces Group provided the initial training.[18] Combined Security Transition Command–Afghanistan, including Task Force Phoenix, then took over the bulk of ANA training. New Afghan recruits received training on basic rifle marksmanship; platoon and company-level tactics; use of heavy weapons; and engineering, scout, and medical skills. Desertion rates were initially high; Afghanistan's 1st Battalion had a desertion rate of approximately 50 percent. But the rate eventually dropped to 10 percent per month by the summer of 2003, between 2 percent and 3 percent per month by 2004, and 1.25 percent per month by 2006.[19]

Several units of the ANA were deployed throughout Afghanistan to conduct combat operations against Taliban and other insurgent forces and to oversee the cantonment of heavy weapons. In July 2003, for example, the ANA launched Operation Warrior Sweep with U.S. forces in the Paktia Province against Taliban and al Qaeda forces. This was followed in November 2003 by Operation Mountain Resolve in the Nuristan and Kunar provinces. The ANA deployed outside of Kabul to stem interfactional fighting in such areas as Herat and Maimana. During the constitutional *loya jirga* in December 2003, the

---

81, No. 4 (July/August 2002), pp. 44–59; Luke Hill, "NATO to Quit Bosnia, Debates U.S. Proposals," *Jane's Defence Weekly*, Vol. 40, No. 23, December 10, 2003, p. 6.

[18] Anthony Davis, "Kabul's Security Dilemma," *Jane's Defence Weekly*, Vol. 37, No. 24, June 12, 2002, pp. 26–27; Mark Sedra, *Challenging the Warlord Culture: Security Sector Reform in Post-Taliban Afghanistan*, Bonn, Germany: Bonn International Center for Conversion, 2002, pp. 28–30.

[19] Attrition has been caused by a number of factors, such as low pay rates and apparent misunderstandings between ANA recruits and the U.S. military. For example, some Afghan soldiers believed they would be taken to the United States for training. On attrition rates, see Afghanistan Ministry of Defense, *Securing Afghanistan's Future: Accomplishments and the Strategic Path Forward, National Army* (Kabul: Ministry of Defense, 2004); U.S. Department of State, *Capitol Hill Monthly Update, Afghanistan* (Washington, D.C.: U.S. Department of State, June 2004); The White House, *Rebuilding Afghanistan* (Washington, DC: The White House, 2004); author interview with U.S. Department of Defense officials, May 2006.

ANA was deployed in the capital region to enhance security for the delegates. In 2004, the ANA conducted combat operations, such as Operation Princess and Operation Ticonderoga, in a number of provinces in the east and south including Kandahar, Kunar, Uruzgan, Helmand, and Paktia. In other provinces, such as Herat, the government deployed ANA and police forces to patrol roads, secure government and UN buildings, and institute a curfew following the removal of Ismail Khan. In 2005, ANA forces participated in such campaigns as Operation Catania in Kunar Province, which targeted insurgent hideouts prior to the September parliamentary elections.[20]

In 2006, ANA soldiers played a key role in a number of offensive campaigns to kill or capture insurgents, including Operation Mountain Thrust in southern Afghanistan and Operation Mountain Lion in Kunar Province.[21] Soldiers from the 3rd Brigade of the ANA's 203rd Corps fought alongside service members from the Combined Forces Command–Afghanistan's Task Force Spartan, made up of soldiers from the U.S. Army's 3rd Brigade Combat Team of the 10th Mountain Division and 1st Battalion, 3rd Marine Regiment U.S. Marines from Task Force Lava. More than 2,500 ANA and coalition forces were involved in the operation.[22]

Evidence from ANA actions during these operations suggests three conclusions about their competence. First, ANA soldiers were usually tenacious fighters in battle and became more proficient in tactics, techniques, and procedures for fighting counterinsurgency warfare following U.S. and coalition training. This included combat proficiency, crowd control, and humanitarian assistance and civil-action

---

[20] "Fighting in Afghanistan Leaves 40 Insurgents Dead," American Forces Press Service, June 22, 2005.

[21] "Coalition Launches 'Operation Mountain Lion' in Afghanistan," American Forces Press Service, April 12, 2006.

[22] U.S. Air Force F-15Es, A-10s and B-52s provided close air support to troops on the ground engaged in rooting out insurgent sanctuaries and support networks. Royal Air Force GR-7s also provided close air support to coalition troops in contact with enemy forces. U.S. Air Force Global Hawk and Predator aircraft provided intelligence, surveillance, and reconnaissance, while KC-135 and KC-10 aircraft provided refueling support.

projects.[23] Second, they were effective in gathering intelligence about insurgents, their support network, and weapon caches. Third, training from U.S. and coalition forces was a critical factor in improving the ANA's competence.[24]

Despite their increasing competence, however, ANA forces still suffered from a lack of indigenous air support and the absence of a self-sustaining operational budget. They also relied on embedded international forces and U.S. air support during combat. The ANA was under-resourced, which was a major morale factor for soldiers. They had shoddy small arms, and there were numerous reports of soldiers using weapons seized from the Taliban, who some believe were better armed. Many soldiers had little ammunition and few magazines. Numerous ANA units did not have mortars and had few machine guns, few MK-19 grenade machine guns, and no artillery. They had almost no helicopter or fixed-wing transport, as well as no attack aviation. They had no body armor or blast glasses, Kevlar®[25] helmets, up-armored Humvee's, or light armor tracked vehicles with machine gun cupolas and slat armor.[26] This impacted their ability to conduct sustained operations against well-equipped Taliban forces.

Other Afghan Security Forces

There is some evidence that Afghan security forces outside of the ANP and ANA were effective in conducting counterinsurgency operations. These forces included the Afghan National Auxiliary Police and militias trained by U.S. special forces, the CIA, and other coalition governments. But these successes were partly outweighed by growing evi-

---

[23] Major Robert W. Redding, "19ᵗʰ SF Group Utilizes MCA Missions to Train Afghan National Army Battalions," *Special Warfare*, Vol. 17, February 2005, pp. 22–27.

[24] Some have argued that the emphasis on quality has a high price tag. For example, a World Bank study concluded, "The ANA salary structure, determined apparently without reference to fiscal constraints or pay elsewhere in the civil service, has set a precedent which the police and other sectors aspire to and which will be fiscally costly." World Bank, *Afghanistan: Managing Public Finances for Development* (Washington, D.C.: World Bank, 2005), p. 24.

[25] Kevlar® is a registered trademark of the DuPont Corporation.

[26] McCaffrey, "Trip to Afghanistan and Pakistan."

dence that these local militia forces weakened the power of the central government.

The Afghan National Auxiliary Police was established in 2006 because of the low quality of the ANP, the inadequate number of ANA soldiers in the south, and the growing level of insurgent violence in the south and east. These developments led to calls within the Afghan government and NATO for a paramilitary police force equipped and trained to fight the Taliban. Some also supported the creation of an auxiliary police force as a way to lure young fighters from local villages away from the Taliban. The auxiliary police force was designed to secure static checkpoints, provide community policing, and deploy with NATO military forces during operations. Recruits were chosen by local tribal elders, and many worked for local warlords. Field reports from the south indicated that they were competent in battle.[27] However, there were several concerns with the auxiliary force. One was that they were a legalized militia more loyal to local warlords than to the central government. Second was suspicion that Taliban members infiltrated the auxiliary police force. The Afghan government and NATO targeted fighters that could—or perhaps already had—fought for the Taliban but were not ideologically committed to Deobandism or jihad against the West. This strategy risked opening the door to Taliban sympathizers. Third was the minimal training that recruits received. Auxiliary police recruits went through a two-week training course, which included lessons on the Afghan constitution, human rights, the use of weapons, and basic police tactics. At the end of the course, the recruits were given an AK-47 gun and sent to their home districts. This training was not sufficient.

Besides the auxiliary police, the U.S. and other coalition forces worked with or established a number of local militias. These militias performed a wide range of missions, including providing base camp security, participating in direct action operations, helping to hunt high-value targets, conducting counternarcotics operations, and providing border security. In Paktika Province, for example, a single squad-size team of U.S. soldiers recruited a company-sized element of Afghans

---

[27] Author interview with NATO and Afghan officials, Afghanistan, January 2007.

to conduct combat operations and stabilization missions; they also recruited another similarly sized unit that provided force protection at remote fire bases from which the U.S. soldiers operated.[28] These units occupied the same areas of operation as the ANA and had a similar mission. But they faced fewer organizational problems, and desertion was practically nonexistent. Soldiers were paid well and regularly, a striking contrast from ANA and ANP forces. Incompetent leaders and undisciplined soldiers were often quickly and severely punished. The best soldiers could expect to move up to positions of increasing responsibility and pay based on fair criteria. As one U.S. military assessment concluded, Afghan militia forces "led every mounted patrol and most major operations," partly because "they knew the ground better and could more easily spot something that was out of place or suspicious."[29] Such forces were often used for the outer perimeter of cordon-and-search operations. They surrounded villages and stopped insurgents from escaping. In several operations, such as the Battle of Deh Chopan in August 2003, they were critical in providing intelligence and the bulk of the maneuver force.[30] As discussed below, however, the U.S. government's use of militia forces also weakened the central government and strengthened local warlords.

National Directorate for Security

There is little unclassified data on the activities and performance of Afghanistan's intelligence agency, the National Directorate for Security (NDS). But its role is critical in counterinsurgency operations. As David Galula wrote in his book *Counterinsurgency Warfare*, gathering intelligence in a counterinsurgency campaign poses a dilemma:

---

[28] Fire bases are military encampments designed to provide support to infantry operating in areas beyond the normal range of major base camps.

[29] Buffaloe, *Conventional Forces in Low-Intensity Conflict*, p. 12.

[30] For a first-hand account of the Battle for Deh Chopan, see Michael McInerney, "The Battle for Deh Chopan, Part 1," *Soldier of Fortune*, August 2004; Michael McInerney, "The Battle for Deh Chopan, Part 2," *Soldier of Fortune*, September 2004.

> Intelligence is the principal source of information on guerrillas,
> and intelligence has to come from the population, but the popu-
> lation will not talk unless it feels safe, and it does not feel safe
> until the insurgent's power has been broken.[31]

In Afghanistan, this posed a challenge for NDS activities in rural
areas in which the Taliban enjoyed a support base. In spite of these
challenges, NDS agents became increasingly involved in intelligence
collection and psychological operations in rural areas of the south and
east, including the use of informant networks. One key target was
mosques. Since the Taliban was successful in co-opting some mullahs,
the NDS focused on influencing "nationalist" mullahs supportive of
the Afghan government or, at a minimum, opposed to the Taliban.
As NDS director Amrullah Saleh argued, "We should put our weight
behind the nationalist ones and not allow the militant or fanatic ones
to take over. This is only possible if we keep the nationalist ones on our
pay-rolls."[32] The NDS also enjoyed some counterinsurgency success in
urban cities, such as Kabul. In late 2006, for example, NDS agents
infiltrated and took down several suicide terrorism cells in Kabul with
assistance from Western intelligence agencies.[33]

Governance

Good governance is critical to effective counterinsurgency operations
over the long run, because it helps to ensure sustained popular support
for the government. As an NDS report concluded, "The first requirement of countering Taliban at the village level requires good governance, honest and competent leaders leading the institutions."[34] Three
factors have undermined good governance in Afghanistan: the power

---

[31] Galula, *Counterinsurgency Warfare*, p. 72.

[32] Saleh, *Strategy of Insurgents and Terrorists in Afghanistan*, p. 10.

[33] Interview with Western government officials, Afghanistan, January 2007.

[34] Saleh, *Strategy of Insurgents and Terrorists in Afghanistan*, p. 10.

of warlords and tribal militias; the drug trade, which has contributed to pervasive corruption; and an ineffective justice system.

Warlords and Tribal Militias

The 2004 presidential elections and 2005 parliamentary elections established a democratically elected central government in Afghanistan. However, they did not create a *strong* central government. Indeed, the weakness of Afghanistan's central government and the role of regional warlords and tribal militias posed a significant challenge to the counterinsurgency campaign. As the Afghanistan National Security Council's National Threat Assessment concluded,

> Non-statutory armed forces and their commanders pose a direct threat to the national security of Afghanistan. They are the principal obstacle to the expansion of the rule of law into the provinces and thus the achievement of the social economic goals that the people of Afghanistan their Government, supported by the International Community, to deliver.[35]

Afghanistan has historically lacked a strong central government. But the reliance of U.S.-led coalition forces on warlords contributed to the empowerment of factional commanders and the weakness of the central government.[36] This began in 2001, when the U.S. military and CIA cooperated with Northern Alliance and other military forces to overthrow the Taliban regime. And it continued over the course of the insurgency. But the assistance supplied to warlords in the south and east by U.S. and coalition forces as part of Operation Enduring Freedom was a double-edged sword. While some cooperation may have been useful and necessary to combat insurgents—especially in the early stages of Operation Enduring Freedom—it also weakened the central government by increasing the power of warlords. The central government made an attempt to reduce the power of warlord-governors

---

[35] Afghanistan National Security Council, *National Threat Assessment* (Kabul: Afghanistan National Security Council, 2005), p. 3. Also see Afghanistan Ministry of Defense, *The National Military Strategy* (Kabul: Afghanistan Ministry of Defense, October 2005).

[36] Ali Jalali, "The Future of Afghanistan," p. 5.

by reassigning them away from their geographic power base, but their networks continued to influence provincial and district-level administration.[37] Indeed, warlords and regional commanders remained strong throughout the country.[38]

Drug Trade

The growth of the drug trade has been well documented. Perhaps its most significant impact was the growth of corruption that paralleled its rise. Afghanistan has a relatively short history as a major opium producer dating back to the 1980s. Criminal organizations, the Taliban, and warlords have historically used drug money to help fund their military campaigns and earn a profit. In 1997, the UN and the United States estimated that 96 percent of Afghan heroin came from areas under Taliban control. The Taliban expanded the area available for opium poppy production, as well as increased trade and transport routes through such neighboring countries as Pakistan.[39] In July 2000, Mullah Omar banned the cultivation—though not the trafficking—of opium poppy, which the Taliban effectively enforced. The ban caused a temporary decrease in the cultivation and production of opium poppy in 2001. But the damage had been done, as Afghanistan had already become a major producer of opium poppy. It was the world's largest producer of poppy during the Taliban rule, when the country was the source of 70 percent of global illicit poppy.

The cultivation, production, and trade in opium poppy increased over the course of the counterinsurgency campaign. Acreage cultivation figures are difficult to estimate, but UN data suggest that the

---

[37] Several warlords were reassigned as provincial governors, such as: Sher Mohammad Akondzada of Helmand (2005), Ismail Khan of Herat (2004), Gul Agha of Kandahar (2004), Haji Din Mohammad of Nangarhar, Mohammad Ibrahim of Ghor (2004), Gul Ahmad of Badghis (2003), and Syed Amin of Badakshan (2003).

[38] MAJ Andrew M. Roe, "To Create a Stable Afghanistan," *Military Review*, November-December 2005, pp. 20–26. On the problem of warlord militias, see *Security Sector Reform: Disbandment of Illegal Armed Groups Programme (DIAG) and Disarmament, Demobilisation, and Reintegration Programme (DDR)* (Kabul: Government of Afghanistan, October 2005).

[39] Rashid, *Taliban*, pp. 119–120.

drug trade remained one of Afghanistan's most serious challenges.[40] As Figure 5.1 illustrates, poppy cultivation rose from approximately 74,045 hectares in 2002 to 131,000 hectares in 2004. It dipped slightly to 104,000 in 2005, and then rose again to 165,000 hectares in 2006 and 193,000 hectares in 2007. Afghanistan's share of opium poppy production stood at 93 percent of the world total in 2007.[41]

Figure 5.1
Opium Poppy Cultivation, 1986–2007

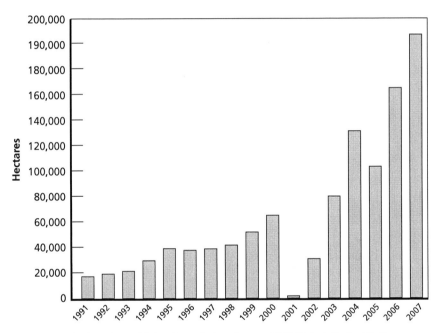

SOURCE: United Nations, *Afghanistan Opium Survey 2007* (Kabul: United Nations Office on Drugs and Crime, 2007); United Nations, *Afghanistan: Opium Survey 2005* (Kabul: United Nations Office on Drugs and Crime, 2005).
RAND *MG595-5.1*

---

[40] Author interview with United Nations officials, Kabul, Afghanistan, November and December 2005.

[41] United Nations, *Afghanistan Opium Survey 2007* (Kabul: United Nations Office on Drugs and Crime, 2007); United Nations, *Afghanistan Opium Survey 2006* (Kabul: United Nations Office on Drugs and Crime, 2006); United Nations, *Afghanistan: Opium Survey 2005* (Kabul: United Nations Office on Drugs and Crime, 2005).

Laboratories in Afghanistan converted opium into morphine base, white heroin, or one of several grades of brown heroin. Afghanistan produced no essential or precursor chemicals for the conversion of opium into morphine base. Acetic anhydride, which is the most commonly used acetylating agent in heroin processing, was smuggled into Afghanistan from Pakistan, India, Central Asia, China, and Europe. The largest processing labs were primarily located in Badakshan, Nangarhar, and Helmand.[42] Most of the opiates produced in Afghanistan were smuggled to markets in the West, although some was consumed in Afghanistan as both opium and heroin. Afghan heroin was trafficked via many routes, with traffickers adjusting smuggling routes based on law enforcement and political actions. Traffickers in Afghanistan primarily relied on vehicles and overland routes to move drug shipments out of the country. Illicit drug convoys transited southern and western Pakistan, while smaller shipments of heroin were sent through the frontier provinces to Karachi for onward shipment. Most of the Afghan heroin that made its way to the United States originated in such provinces as Nangarhar and transited Pakistan.[43]

The drug trade has traditionally been a source of revenue for warlords, insurgents, and criminal organizations, as well as members of the Afghan government.[44] In June 2005, for example, the U.S. Drug Enforcement Agency and Afghan Counternarcotics Police raided the offices of Sher Mohammed Akhundzada, governor of Helmand Province. They found over nine metric tons of opium stashed there.[45] It is difficult to assess why cultivation numbers dropped briefly in 2005 before rising again in 2006, but most of the reduction appears to have been the result of persuasion and coercion. The UN Office on Drugs and Crime concluded that "the Government of Afghanistan ordered

---

[42] Correspondence with former Afghan Minister of Interior Ali Jalali, September 5, 2006.

[43] Tandy, *Statement.*

[44] Author interview with European government officials, 2006. Also see Barnett R. Rubin and Andrea Armstrong, "Regional Issues in the Reconstruction of Afghanistan," *World Policy Journal*, Vol. XX, No. 1 (Spring 2003), p. 34.

[45] Tandy, *Statement.*

provincial governors to eradicate opium poppy fields."[46] For example, the governor of Nangarhar played a role in reducing cultivation percent from 2004 to 2005 using prevention techniques and intimidation by Afghan police. The U.S. government also provided economic assistance. Almost three-fourths of the eradication (72 percent) took place in Nangarhar and Helmand provinces, where, in 2004, poppy cultivation was ranked highest in the nation.[47]

Viable and sustainable income-generation programs need to be established to support eradication efforts and to help convince some farmers not to cultivate. In some locations, eradication was accompanied by alternative livelihood programs, material support, and significant political persuasion. The provinces where declines in cultivation were most striking in 2005 (Nangarhar—96 percent, Badakshan—53 percent) or where cultivation remained relatively stable (Helmand—10 percent) were the three provinces that received the largest contributions for alternative development. Nangarhar received $70.1 million in assistance, and Badakshan and Helmand received $47.3 million and $55.7 million respectively.[48] In spite of these temporary successes, the drug trade will remain a major challenge for the foreseeable future.[49]

Justice System

Afghanistan's justice system faced severe problems. Measuring the effectiveness of the justice system is problematic in the absence of reliable data. However, World Bank data suggest that Afghanistan's rule of law was one of the least effective—if not the least effective—in the world. These data measure the extent to which populations have confidence

---

[46] United Nations, *Afghanistan: Opium Survey 2005*, p. iii.

[47] Author interview with Doug Wankel, Director of the Office of Drug Control, U.S. Embassy, Kabul, November 2005; United Nations, *Afghanistan: Opium Survey 2005*, p. iii.

[48] United Nations, *Afghanistan: Opium Survey 2005*, pp. iii–iv.

[49] United Nations, *Afghanistan: Opium Survey 2005*; United Nations, *Afghanistan: Opium Survey 2004* (Vienna: United Nations Office on Drugs and Crime, 2004); United Nations, *The Opium Economy in Afghanistan: An International Problem*, New York: United Nations Office on Drugs and Crime, 2003; United Nations, *Afghanistan: Opium Survey 2003* (Vienna: United Nations Office on Drugs and Crime, 2003), pp. 1–10; Rubin, *Road to Ruin*.

in, and abide by, the rules of society. They include perceptions of the incidence of crime, the effectiveness and predictability of the judiciary, and the enforceability of contracts.[50] Figure 5.2 illustrates Afghanistan's rule of law in comparison to other countries in the region. The data show that Afghanistan's justice system started from a low base. When the United States helped overthrow the Taliban regime in 2001, Afghanistan had the lowest-ranking justice system in the world, and it did not significantly improve over the course of reconstruction efforts. In comparison to other countries in the region—such as Iran, Pakistan, Russia, Tajikistan, Turkmenistan, and Uzbekistan—Afghanistan's justice system was one of the least effective.

Figure 5.2
Afghanistan's Rule of Law, 1996–2006

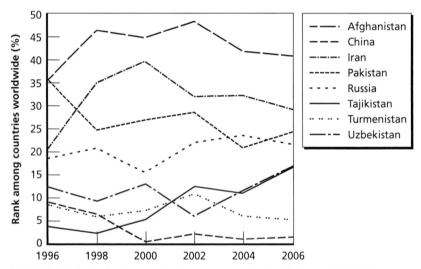

SOURCE: World Bank, *Aggregate Governance Indicators Dataset, 1996–2006* (Washington, D.C.: World Bank, 2007).
RAND MG595-5.2

---

[50] Daniel Kaufmann, Aart Kraay, and Massimo Mastruzzi, *Governance Matters V: Aggregate and Individual Governance Indicators for 1996–2005*, Washington, D.C.: World Bank, 2006, p. 4.

There have been several challenges to improving Afghanistan's justice system, all of which have severely impacted the efficacy of the counterinsurgency campaign. First, the central government's inability to decrease the power of warlords and exert control over the country impacted justice sector reform. Warlord commanders, who were allowed to maintain de facto control over areas seized following the overthrow of the Taliban regime, established authority over local courts. The factional control of courts led to intimidation of centrally appointed judges. Second, the Afghan government's inability and unwillingness to address widespread and deep-rooted corruption decreased the effectiveness of the justice system. Corruption has long been endemic in the justice system, partly because unqualified personnel loyal to various factions are sometimes installed as court officials. The supreme court and attorney general's office were accused of significant corruption.[51] The World Bank concluded that Afghanistan was one of the most corrupt governments in the world.[52] A corrupt judiciary is a serious impediment to the success of a counterinsurgency campaign. It threatens to further undermine good governance and popular support, and it cripples the legal and institutional mechanism necessary to prosecute insurgents and criminals.

Conclusion

The quality of Afghan forces and governance capacity was mixed in Afghanistan. As noted in Chapter Two, the effectiveness of counterinsurgency operations can be directly correlated with the competence of indigenous security forces and the strength of governance. This has been a recurring lesson across insurgencies since at least 1945. The recommendations put forth in Chapter Seven will explore in more depth U.S. capabilities that might be useful in improving indigenous capabilities.

---

[51] Author interview with Deputy Minister of Justice Mohammad Qasim Hashimzai, Kabul, Afghanistan, June 26, 2004; Rama Mani, *Ending Impunity and Building Justice in Afghanistan* (Kabul: Afghanistan Research and Evaluation Unit, 2003), p. 2.

[52] Kaufmann, Kraay, and Mastruzzi, *Governance Matters V,* pp. 113–115.

CHAPTER SIX

U.S. and Coalition Forces

This chapter examines U.S. and coalition forces in Afghanistan. The United States initially played the lead role in the counterinsurgency campaign, though command and control over most international forces shifted to NATO in late 2006. Whereas Chapters Four and Five focused on such critical factors as the role of external support for insurgents, the legitimacy and capacity of the indigenous government, and the quality of local forces, this chapter examines six areas related directly to U.S. efforts:

- building indigenous capacity
- direct action against insurgents
- intelligence
- information operations
- coalition operations
- civil-military affairs.

It argues that U.S. and coalition efforts had mixed results. First, U.S. counterinsurgency capabilities were most effective when they leveraged the Afghan government and indigenous forces. Provincial Reconstruction Teams (PRTs) and Team Village missions were effective in rebuilding some infrastructure and garnering popular support. Special forces were particularly effective in conducting kinetic and nonkinetic operations with Afghan forces, and building indigenous capacity. Human intelligence (HUMINT) and civil-military operations were often most effective when conducted using indigenous forces. And indigenous leaders, such as Muslim clerics and tribal elders, were helpful in conducting information operations. Second, U.S. forces were least effective when they acted unilaterally and failed to leverage the indigenous government and its forces. In some areas, such as inside Pakistan, there were no sustained counterinsurgency operations by Pakistani or international forces. The ability to integrate indigenous forces into operations and to conduct civil-military operations during the counterinsurgency campaign also varied depending on whether U.S. conventional or unconventional ground forces were involved. Again, special forces were particularly effective.

Building Indigenous Capacity

Chapter Five focused on the quality and legitimacy of Afghan security forces, and argued that there were notable problems with the ANP. This section looks more specifically at the U.S. "light footprint" approach and the use of indigenous forces during operations. Metaphorically, counterinsurgency is about teaching people to fish, not about doing it for them. One of the most significant lessons from U.S. and coalition experiences in Afghanistan is the importance of advising, training, and assisting the host nation's ministries and security forces—and helping shift popular support from the insurgents to the host nation.[1] For insurgent groups, popular support is an overriding strategic objective. As Mao Tse-tung argued, "The richest source of power to wage war lies in the masses of the people."[2] Counterinsurgency operations must separate insurgents from their support base.

The United States adopted a two-pronged strategy for counterinsurgency operations in Afghanistan. The first part involved the establishment of an indigenous (rather than international) government. As noted in Chapter Five, public opinion polls conducted in 2003 and 2004 demonstrated that roughly 85 percent of Afghans interviewed had a "very favorable" or "somewhat favorable" view of Hamid Karzai, the

---

[1] Herd, *World War III.*

[2] Mao Tse-Tung, *Selected Military Writings of Mao Tse-Tung* (Peking, China: Foreign Languages Press, 1963), p. 260.

first post-Taliban president of Afghanistan.[3] The United States and its allies fostered the rapid creation of a national Afghan government that included representatives from major ethnic groups. This new government began to form even before major combat operations terminated. On December 5, 2001, Afghan delegates executed the Bonn Agreement, which established an interim administration under Hamid Karzai and called for an Emergency Loya Jirga to establish a transitional authority. Karzai became head of state in June 2002 and president in October 2004, following the country's first national election. The establishment of an interim indigenous—rather than international—government gave the population a sense of ownership. As Zalmay Khalilzad, special envoy and former U.S. ambassador to Afghanistan, argued, "The fact that the Afghans played a key role in their own liberation gave them a sense of dignity and ownership of their destiny."[4]

The second prong of the U.S. approach involved maintaining a light military footprint in Afghanistan. The United States did not invade Afghanistan with large numbers of forces. U.S. officials adopted a light-footprint approach for several reasons: They wanted to prevent large-scale resistance similar to what the Soviet Union encountered in the 1980s; they believed that small numbers of ground troops and airpower were sufficient to establish security; and they were deeply reluctant to become involved in nation-building.[5] Indeed, several great powers throughout history have been defeated in Afghanistan, including the forces of Alexander the Great, Great Britain, and the Soviet Union.

---

[3] The Asia Foundation, *Voter Education Planning Survey: Afghanistan 2004 National Elections* (Kabul: The Asia Foundation, 2004), pp. 107–108.

[4] Zalmay Khalilzad, "How to Nation-Build: Ten Lessons from Afghanistan," *National Interest*, No. 80 (Summer 2005), pp. 20–21.

[5] Author interview with Ambassador James Dobbins, former U.S. envoy to Afghanistan, Washington, D.C., September 21, 2004 and November 15, 2006; Richard Clarke, *Against All Enemies: Inside America's War on Terror* (New York: Free Press, 2004); U.S. Senate, *Afghanistan Stabilization and Reconstruction: A Status Report—Hearing Before the Committee on Foreign Relations*, S.Hrg. 108-460, January 27, 2004, pp. 14, 17–18; Seymour M. Hersh, "The Other War: Why Bush's Afghanistan Problem Won't Go Away," *The New Yorker*, April 12, 2004; United Nations, *Report of the Secretary-General on the Situation in Afghanistan and Its Implications for International Peace and Security*, UN doc A/56/875-S/2002/278, March 18, 2002, para. 98.

U.S. GEN Tommy Franks, who put together the operational concept for Afghanistan in 2001, argued that, after major combat ended, "our footprint had to be small, for both military and geopolitical reasons. I envisioned a total of about 10,000 American soldiers, airmen, special operators, and helicopter assault crews, along with robust in-country close air support."[6]

This strategy worked well during the overthrow of the Taliban government. But it had two drawbacks for the stability operation phase, which began after the overthrow of the Taliban regime. First, there were too few U.S. and Afghan government forces to stabilize the country.[7] Afghan militia forces had to fill the security vacuum, which, in the long run, undermined the power of the central government. Figure 6.1 illustrates the per capita level of external forces in 17 stability operations since World War II. The United States had one of the lowest per capita levels in Afghanistan among all these operations, which even included a number of UN operations in Africa and Asia. This created a challenge in targeting Taliban, Hezb-i-Islami, and al Qaeda insurgents in the early stages of the insurgency—such as 2002 and 2003—since there were virtually no trained and legitimate Afghan military and police. Nor were there sufficient forces to secure Afghanistan's borders. Insurgent forces benefited from porous borders along the Afghanistan-Pakistan frontier, as well as from assistance from sources in Pakistan and the Muslim world channeled through Pashtun tribesmen in the border region.

Second, this strategy drew the wrong lesson from the Soviet experience. The key lesson from the Soviet experience was not the number of Soviet forces deployed, but rather how they were used. The Soviets fought the wrong war; they fought a conventional war against an unconventional opponent. As one of the most comprehensive studies of Soviet combat tactics in Afghanistan concluded, "The Soviet Army that marched into Afghanistan was trained to fight within the context of a theater war against a modern enemy who would obligingly occupy

---

[6] GEN Tommy Franks, *American Soldier* (New York: HarperCollins, 2004), p. 324.

[7] Seth G. Jones, "Averting Failure in Afghanistan," *Survival*, Vol. 48, No. 1 (Spring 2006), pp. 111–128.

Figure 6.1
Peak Military Presence per Capita

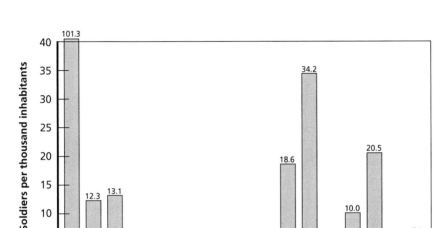

SOURCE: James Dobbins, Seth G. Jones, Keith Crane, Andrew Rathmell, Brett Steele, Richard Teltschik, and Anga Timilsini, *The UN's Role in Nation-Building, from the Congo to Iraq*, Santa Monica, Calif.: RAND Corporation, MG-304-RC, 2005, p. 228.
RAND *MG595-6.1*

defensive positions stretching across the northern European plain." The Soviets used massed artillery, tanks, and ground forces to destroy acres of defense positions, and "Soviet tactics and equipment were designed solely to operate within the context of this massive strategic operation."[8] The Soviets fundamentally misunderstood the nature of counterinsurgency warfare. They terrorized the population instead of working to win the people over to the government's side. The United States, Pakistan, Saudi Arabia, and other governments exploited this resentment by providing military and financial assistance to the mujahideen.[9]

---

[8] Grau, *The Bear Went Over the Mountain*, p. 201.

[9] Coll, *Ghost Wars*, pp. 131, 167, 202.

This brings up an important dilemma. A lead indigenous effort is critical over the long run for successful counterinsurgency operations. Even if tactically successful, a unilateral external operation may ultimately lead to failure by unseating the very indigenous capability that the external actor is trying to build. But what if there is no competent government force in the early stages of an insurgency? In the Afghan case, there were no Afghan army forces and no trained police. While there are no ideal options in these situations, the most effective strategy may be to (a) work with those legitimate indigenous forces (especially police) that exist; (b) effectively train and mentor them as quickly as possible; and (c) temporarily back-fill indigenous security forces with sufficient numbers of U.S. and other international forces to accomplish key security tasks. These tasks include patrolling streets and villages, monitoring borders, and protecting critical infrastructure. Higher per capita levels of U.S. and coalition military and police forces might have been useful in the immediate aftermath of the Taliban's overthrow. Preparations for the war in Iraq made this increasingly difficult, because U.S. troops were needed for combat operations there.

Once counterinsurgency operations began, the United States experienced varying degrees of success in working with indigenous forces. Special forces were particularly successful at integrating ANA, ANP, and Afghan militia forces in virtually all aspects of combat and civil-military operations. Indigenous forces were involved in conducting strike operations; interdicting enemy forces along the border; participating in reconstruction efforts; and gathering intelligence.[10]

Some conventional units were also successful in working with indigenous forces. For example, forces from the U.S. Army's 82nd Airborne Division in the Bermel Valley involved Afghan militia forces and ANA forces in mine detection, mounted patrols, intelligence collection, vehicle checkpoints, combat operations, and search-and-seize missions. As one assessment summarized, Afghan forces "led every joint mounted patrol and most major operations," especially since they

---

[10] Combined Joint Special Operations Task Force Afghanistan, *Counterinsurgency Operations in Afghanistan, July to December 2004: Principles of Victory* (Combined Joint Special Operations Task Force Afghanistan, 2005).

"knew the ground better and could more easily spot something that was out of place or suspicious."[11] But others were less successful. During Operation Mountain Sweep, for example, paratroopers from the 82nd Airborne conducted operations unilaterally; their heavy-handed tactics created significant resentment among locals in Khowst Province.[12] U.S. soldiers angered Afghan villagers on numerous occasions out of naiveté of Afghan social and cultural traditions, leading one Afghan government report to conclude,

> It will be difficult for ever for the coalition forces to fully befriend the people. Instead they should try to minimize their contact with the local population and increasingly empower the Afghan forces to do the job. The more they try to be in touch with people the more they will be prone to make cultural mistakes.[13]

Direct Action Against Insurgents

U.S. military forces succeeded to varying degrees in four areas of combat operations: the use of a "clear, hold, and expand" approach; the use of armed reconnaissance and raiding; close air support; and command and control arrangements. Direct action was most effective when the aim was to reduce the use of force, to use force in ways considered legitimate, and to get locals to use force instead of U.S. and coalition forces.

Clear, Hold, and Expand
Counterinsurgents achieve success by destroying insurgent forces and their political organization in a given area over the long run. This involves the permanent isolation of the insurgents from the population.

---

[11] Buffaloe, *Conventional Forces in Low-Intensity Conflict*, p. 12.

[12] Hy S. Rothstein, *Afghanistan and the Troubled Future of Unconventional Warfare* (Annapolis, Md.: Naval Institute Press, 2006), pp. 141–143; MAJ Ron Sargent, "Strategic Scouts for Strategic Corporals," *Military Review*, Vol. 85, No. 2 (March-April 2005), pp. 12–17.

[13] Saleh, *Strategy of Insurgents and Terrorists in Afghanistan*, p. 6.

Ideally, this isolation is not enforced on the population but maintained by and with them.[14]

One of the most successful approaches in Afghanistan (when it was applied) was "clear, hold, and expand." This has also been referred to as an "ink-spot" strategy, in which military forces set up secure zones and then slowly expanded them outward like ink spots on blotting paper. Forces were assigned to contested areas to regain government presence and control and then conducted military and civil-military programs to expand the control and edge out insurgents.[15] The focus was on consolidating and holding ground that was clearly pro-Afghan and procoalition (or at least anti-Taliban); protecting the government and other key resources (such as lines of communication and major cities like Kabul); and deploying coalition counterinsurgency forces to conduct offensive operations in contested areas of Afghanistan, especially in the south and east. The deployment of forces into insurgent areas was designed to deny sanctuary, interdict the border, and expand government and coalition presence. This is discussed in more detail in the next section. U.S. counterinsurgency forces were kept to a bare minimum and supported with civil affairs and psychological operations personnel. A company of infantry has sometimes been provided for area patrolling and to provide security against an immediate threat to the unit. Quick reaction forces in the form of close air support assets or reinforcing units have backed up the outposts whenever insurgent forces have threatened to overrun them.

Clear, hold, and expand forces conducted operations in ever-increasing zones, or ink spots, around their bases. In the first zone, forces tried to target and eliminate the insurgents living within the area. This required living among the local population for long durations to gain its trust and support and then trying to separate the locals from the insurgents. The secondary zone was the transit and support zone for the insurgents. Clear, hold, and expand forces in Afghani-

---

[14] Galula, *Counterinsurgency Warfare*, p. 77.

[15] The clear, hold, and expand section draws extensively from Joseph D. Celeski, *Operationalizing COIN*, JSOU Report 05-2 (Hurlburt Field, Fla.: Joint Special Operations University, 2005).

stan cast a wide net of operations outside their force protection zone to disrupt and interdict insurgent operations. This required patience and discreet intelligence work to ascertain the location of insurgent weapon caches, safe houses, and transit support systems. The outer zone included remote locations or areas where the population was neither friendly nor hostile to the counterinsurgency unit's efforts. Occasional operations were conducted in these areas to show the flag and to keep the population neutral to the idea of supporting the insurgents. Battalion-sized sweeps and clearing operations by conventional forces generally reaped far less than their effort because of the difficulty of finding and fixing elusive insurgents.[16]

However, there were some challenges with clear, hold, and expand. Pakistani forces never attempted sustained clear, hold, and expand operations in the tribal areas or in Baluchistan, especially against high- and middle-level Taliban members. Even in Afghanistan, clear, hold, and expand was limited to small areas of the country, since there were too few U.S., coalition, and Afghan forces to hold and expand large areas.

Armed Reconnaissance and Raiding

An additional lesson from the Afghan counterinsurgency is the need to develop an armed reconnaissance capability and a specialized raiding force. Armed reconnaissance is the patrolling of suspected insurgent areas to glean information on their activities, initiate contact and conduct battle, or confirm that the area is clear. Armed reconnaissance in Afghanistan was accomplished with a variety of platforms and measures. These were tailored for "hunter-killer" type missions—search for, hunt down, gain contact with, and keep contact with insurgents. AC-130 gunships (operating generally at night), tactical unmanned aerial vehicles, and mounted ground reconnaissance patrols all served to accomplish this mission and keep insurgents off balance and disrupt their timing. Tactical unmanned aerial vehicles were somewhat help-

---

[16] Celeski, *Operationalizing COIN.*

ful in assisting reconnaissance, force protection, viewing avenues of approach, and positive target identification.[17]

A specialized raiding force was sometimes required to conduct time-sensitive targeting beyond the scope of conventional forces. These specialized raiding forces took various forms, including counterterrorist units, indigenous strike forces, and specially formed and trained units with personnel drawn from organic forces. In Afghanistan, raiding forces often required dedicated mobility platforms and a high level of access to intelligence assets. The sensor-to-shooter links worked best when noncontributing layers of decisionmakers were removed. The number-one role for these units was to target the insurgent's organizational structure and leadership; they also had a secondary role in the conduct of raids in sanctuaries in which political sensitivities precluded larger operations.[18]

Close Air Support

Close air support provided a significant advantage to small groups of U.S. and Afghan forces operating against insurgents. For example, close air support was extremely effective when U.S. forces encountered unexpectedly strong resistance during Operation Anaconda in the Shah-i-Kot Valley. This lesson may not be applicable to all counterinsurgency operations, especially those conducted in urban areas. Also, in cases in which there is sufficient indigenous air capacity, the United States may choose not to provide close air support. Over the course of the Afghan counterinsurgency, a variety of aircraft provided close air support to U.S. and Afghan forces operating in the border areas. Prominent aircraft used include AH-64 attack helicopters, Spectre AC-130 gunships (which some Afghan insurgents referred to as the "water buffalo"), A-10 and F-14 fighters, and B-52 bombers.[19]

---

[17] COL Bruce Burda, *Operation Enduring Freedom Lessons Learned* (Hurlburt Field, Fla.: Air Force Special Operations Command, 2003).

[18] Celeski, *Operationalizing COIN*.

[19] Bruce R. Pirnie, Alan J. Vick, Adam Grissom, Karl P. Mueller, and David T. Orletsky, *Beyond Close Air Support: Forging a New Air-Ground Partnership* (Santa Monica, Calif.: RAND Corporation, MG-301-AF, 2005).

Due to the effectiveness of close air support, it will be particularly useful for the United States to continue to develop technological capabilities like GPS and Special Operations Forces Laser Acquisition Markers (SOFLAMs), which proved invaluable for ground forces. For much the same reason, it will also be useful to continue to develop new and more sophisticated communication equipment, including advanced receivers and transmitters that link forces in the field to those in other areas, satellite radios carried by combat controllers to call in air strikes, and encrypted high-frequency (HF) radio.[20] As one CIA officer involved in operations in Afghanistan remarked, "The advantage provided by SOFLAMs, smart bombs, laser-guided munitions, Spectre AC-130 gunships, Predator drones, sophisticated communications equipment, etc. tipped the balance militarily."[21]

Command and Control

Perhaps the most significant command and control lesson that can be learned from the counterinsurgency in Afghanistan was the need to decentralize authority down to the small unit level. Command and control worked best when it was flattened out from hierarchical to more horizontal levels. The shorter sensor-to-shooter links were, the better they worked. Quicker and more responsive arrangements for command and control provided flexibility for forces on the battlefield.[22]

This was not always well executed. In December 2001, the commander of Task Force Dagger (essentially the 5th Special Forces Group plus supporting units) was in direct contact with General Franks, the combatant commander. The subordinate elements of Task Force Dagger had the most current and accurate intelligence about the situation on the ground. But this changed when the 10th Mountain Division assumed operations in Afghanistan in March 2002 and again when 18th Airborne Corps took over control of the Afghanistan theater of operations in June 2002. By late 2002, a special forces detach-

---

[20] On the utility of SOFLAMs and communication equipment in Afghanistan, see Berntsen, *Jawbreaker*, pp. 78–79, 83, 134, 266–268.

[21] Berntsten and Pezzullo, *Jawbreaker*, p. 313.

[22] Celeski, *Operationalizing COIN*, p. 83.

ment's request to conduct an operation sometimes had to be processed through six levels of command before being approved. As one general officer put it: "Too much overhead."[23] Tight command and control sometimes paralyzed initiative. Reports from some special forces members in Afghanistan indicated that U.S. forces had to obtain approval from the Combined Joint Task Force headquarters before conducting operations six kilometers beyond their fire bases. In addition, all "named operations"—those other than routine travel—required approval from the Combined Joint Task Force headquarters, which could take as long as 48 hours.[24] The net effect of these restrictions was to impede the flexibility and response of special forces.

The mission profiles of future counterinsurgency operations may require adapting organizational structures in several ways. One is to empower operations at the lowest level. Well-trained, small-unit maneuver is important to success.[25] Afghan insurgent groups frequently dispersed their forces, making them smaller and more difficult to attack. They also used more secure communication, better camouflage, and more effective diversions.[26] U.S. military operations most often succeeded when leaders at the small-unit level had enough leeway, specialized assets, and firepower to engage the population and develop their own intelligence.[27] Indeed, U.S. military doctrine needs to establish far looser and more broadly distributed networks that have a high degree of individual independence and survivability.[28] This means incorporating into counterinsurgency doctrine and training the preparation of company and battalion commanders to lead combined-arms warfare,

---

[23] Quoted in Rothstein, *Afghanistan and the Troubled Future of Unconventional Warfare*, p. 111.

[24] Author interviews with U.S. forces in Afghanistan, Paktika and Paktiya, 2004 and 2005.

[25] U.S. Army Training and Doctrine Command, *Operation Enduring Freedom*, pp. 22–23.

[26] Biddle, *Afghanistan and the Future of Warfare*; Anthony H. Cordesman, *The Ongoing Lessons of Afghanistan: Warfighting, Intelligence, Force Transformation, and Nation Building* (Washington, D.C.: Center for Strategic and International Studies, 2004), pp. 122–123.

[27] Buffaloe, *Conventional Forces in Low-Intensity Conflict*, p. 4.

[28] Cordesman, *The Ongoing Lessons of Afghanistan*.

conduct civil-military operations, and develop and exploit their own intelligence. It also means giving infantry commanders the responsibility, autonomy, and distance from higher headquarters that is now only held by special forces A-team commanders. Commanders must empower small-unit leaders to deal with the challenges encountered during counterinsurgency operations, including the authority to routinely make decisions currently made by battalion and brigade combat team commanders.[29]

Intelligence

British Colonel C. E. Callwell, a military historian, wrote in his book *Small Wars* that "it is a very important feature in the preparation for, and the carrying out of, small wars that the regular forces are often working very much in the dark from the outset . . . What is known technically as 'intelligence' is defective, and unavoidably so."[30] Intelligence is the principal source of information on insurgents, and has usually come from the population.[31] U.S. military and intelligence forces used a variety of ways to identify insurgents: signals intelligence (SIGINT); scouts; long-range reconnaissance detachments; unmanned aerial vehicles; remote battlefield sensor systems; Q-36 Firefinder radar; Joint Land Attack Cruise Missile Defense Elevated Netted Sensor forward-looking infrared systems; and an assortment of HUMINT sources.

The U.S. military's experience in Afghanistan illustrates two key lessons: HUMINT usually provides the majority of actionable intelligence, especially at the tactical level; and civil-military operations can be a useful way to gather intelligence.

HUMINT was a critical facet of counterinsurgency operations in Afghanistan, and Afghan intelligence and security forces were vital

---

[29] U.S. Army Training and Doctrine Command, *Observations and Lessons Learned: Task Force Devil, 1st Brigade Combat Team, 82 Airborne Division* (Fort Leavenworth, Kan.: U.S. Army Training and Doctrine Command, January 2004), p. 2.

[30] Callwell, *Small Wars*, p. 43.

[31] Galula, *Counterinsurgency Warfare*, p. 72.

conduct civil-military operations, and develop and exploit their own intelligence. It also means giving infantry commanders the responsibility, autonomy, and distance from higher headquarters that is now only held by special forces A-team commanders. Commanders must empower small-unit leaders to deal with the challenges encountered during counterinsurgency operations, including the authority to routinely make decisions currently made by battalion and brigade combat team commanders.[29]

Intelligence

British Colonel C. E. Callwell, a military historian, wrote in his book *Small Wars* that "it is a very important feature in the preparation for and the carrying out of, small wars that the regular forces are often working very much in the dark from the outset . . . What is known technically as 'intelligence' is defective, and unavoidably so."[30] Intelligence is the principal source of information on insurgents, and has usually come from the population.[31] U.S. military and intelligence forces use a variety of ways to identify insurgents: signals intelligence (SIGINT) scouts; long-range reconnaissance detachments; unmanned aerial vehicles; remote battlefield sensor systems; Q-36 Firefinder radar; Joint Land Attack Cruise Missile Defense Elevated Netted Sensor forward-looking infrared systems; and an assortment of HUMINT sources.

The U.S. military's experience in Afghanistan illustrates two key lessons: HUMINT usually provides the majority of actionable intelligence, especially at the tactical level; and civil-military operations can be a useful way to gather intelligence.

HUMINT was a critical facet of counterinsurgency operations in Afghanistan, and Afghan intelligence and security forces were vital

---

[29] U.S. Army Training and Doctrine Command, *Observations and Lessons Learned: Task Force Devil, 1st Brigade Combat Team, 82 Airborne Division* (Fort Leavenworth, Kan.: U.S. Army Training and Doctrine Command, January 2004), p. 2.

[30] Callwell, *Small Wars*, p. 43.

[31] Galula, *Counterinsurgency Warfare*, p. 72.

Information Operations

How can insurgents hope to succeed? They need to find support among the population. This support may range from active participation in the struggle to passive approval of it. The first basic need for an insurgent aiming at more than simply making trouble is an attractive cause. With a popular cause, the insurgent has a formidable, if intangible, asset that he can progressively transform into concrete strength.[35]

Religion has been a significant part of insurgent rhetoric focused on gaining popular support. As Mullah Dadullah, a Taliban military commander killed in 2007, argued,

> We are not fighting here for Afghanistan, but we are fighting for all Muslims everywhere and also the Mujahideen in Iraq. The infidels attacked Muslim lands and it is a must that every Muslim should support his Muslim brothers.[36]

This argument was echoed by other insurgents, such as former Taliban spokesman Mofti Latifollah Hakimi: "The issue of Afghanistan is connected with the ongoing war between Islam and blasphemy in the world. Mullah Mohammad is representing a huge umma [Muslim community], and a large nation is behind him."[37] Portraying the United States and other Western countries as dedicated to the destruction of Islam was a critical part of this effort. As Mullah Dadullah argued, "God be praised, we now are aware of much of the U.S. plans. We know their target, which is within the general aim of wiping out Islam in this region."[38] The Taliban used young Pakistan-trained mullahs to glorify their cause in mosques in the east and south.

These Taliban efforts had mixed success in Afghanistan, with more success in Pakistan. Since mosques have historically served as

---

[35] Galula, *Counterinsurgency Warfare*, pp. 18–19.

[36] Interview with Mullah Dadullah, Al Jazeera TV, February 2006. Also see, for example, "Taliban Spokesman Condemns Afghan Parliament as 'Illegitimate,'" Sherberghan Aina TV, December 19, 2005.

[37] "Spokesman Rejects Afghan Government's Amnesty Offer for Taliban Leader."

[38] "Taliban Military Chief Threatens to Kill U.S. Captives."

a tipping point for major political upheavals in Afghanistan, Afghan government officials focused on mosques. As one NDS report noted,

> There are 107 mosques in the city of Kandahar out of which 11 are preaching anti-government themes. Our approach is to have all the pro-government mosques incorporated with the process and work on the eleven anti-government ones to change their attitude or else stop their propaganda and leave the area.[39]

Another major factor was the public campaign by Afghan religious figures. For example, in July 2005, the Ulema Council of Afghanistan called on the Taliban to abandon violence and support the Afghan government in the name of Islam. It also called on the religious scholars of neighboring countries—including Pakistan—to help counter the activities and ideology of the Taliban and other insurgent organizations.[40] A number of Afghan Islamic clerics publicly supported the Afghan government and called the jihad un-Islamic.[41] Moreover, the Ulema Council and some Afghan ulema issued fatwas, or religious decrees, that unambiguously oppose suicide bombing. They argued that suicide bombing does not lead to an eternal life in paradise, does not permit martyrs to see the face of Allah, and does not allow martyrs to have the company of 72 maidens in paradise. These efforts to counter Taliban propaganda were made easier by the populations' lingering resentment of the Taliban. Afghan support for the Taliban through 2007 was low. In one public opinion poll, for example, only 13 percent of Afghans had a favorable view of the Taliban.[42]

Insurgent groups also failed to successfully use ethnicity—especially Pashtun nationalism—to gain support, in spite of a Taliban information campaign that included dropping leaflets, delivering night

---

[39] Saleh, *Strategy of Insurgents and Terrorists in Afghanistan*, p. 8.

[40] "Religious Scholars Call on Taliban to Abandon Violence," Pajhwok Afghan News, July 28, 2005.

[41] "Taliban Claim Killing of Pro-Government Religious Scholars in Helmand," Afghan Islamic Press, July 13, 2005.

[42] The Asia Foundation, *Voter Education Planning Survey*, pp. 107–108.

letters, and launching a radio station.[43] This reflects successful U.S. and Afghan government efforts to balance representation in the government among the country's ethnic groups. The Taliban has long drawn its membership from the Pashtuns.[44] In the immediate aftermath of the December 2001 Bonn Conference, the Tajiks and Uzbeks (who comprised the Northern Alliance) filled key government positions with their own personnel. They also took control of the intelligence service and ministries of defense, interior, and foreign affairs. This meant that there was a significant absence of representation from the Pashtuns, as well as the Shia Muslim Hazaras from the center. One notable exception was the decision to name Hamid Karzai, a Pashtun, as interim president. By late 2003, however, the U.S. government and President Karzai made a concerted push to redress the ethnic balance at the levels of minister and deputy minister. For example, President Karzai appointed Ali Jalali, a Pashtun, as Minister of Interior, who began to appoint ethnically diverse governors and police chiefs.[45] In addition, the Afghanistan and Pakistan governments encouraged the organization of traditional Pashtun councils (jirgas) to solicit the aid of local tribes in fighting extremists. This included the use of tribal militias to help quell the growing power of the Taliban in Pashtun areas. The expansion of good governance and strong state institutions is probably the most effective long-term solution.

In sum, indigenous actors were most effective in conducting information operations. The U.S. and other coalition governments tried to implement information operations to degrade insurgent decisionmak-

---

[43] "Taliban Launch Pirate Radio Station in Afghanistan," Agence France Presse, April 18, 2005.

[44] There are no accurate statistics on ethnicity. Based on electoral results, however, Pashtuns appear to account for approximately 50 percent of Afghans. On Pashtuns and the Taliban, see Roy, *Islam and Resistance in Afghanistan*; Rashid, *Taliban*; Maley, *Fundamentalism Reborn?*

[45] S. Frederick Starr, "Sovereignty and Legitimacy in Afghan Nation-Building," in Francis Fukuyama, ed., *Nation-Building: Beyond Afghanistan and Iraq* (Baltimore, Md.: Johns Hopkins University Press, 2006), pp. 107–124; S. Frederick Starr, *U.S. Afghanistan Policy: It's Working* (Washington, D.C.: Central Asia-Caucasus Institute, Johns Hopkins University, 2004).

ing and recruitment. Key objectives were to deter, discourage, and dissuade insurgents by disrupting their unity of command while preserving Afghan and coalition command. They also involved shutting down insurgent communications and networks while protecting Afghan and coalition ones. This included employing five core capabilities: electronic warfare, psychological operations, operations security, military deception, and computer network operations.[46] However, the most successful information operations were from indigenous actors such as religious, tribal, and political leaders—often without U.S. assistance.

Working with Coalition Partners

The U.S. experience working with coalition forces and other international actors was mixed. The counterinsurgency campaign—and security sector reform more broadly—was initially based on a "lead nation" approach. The United States was the lead donor nation for reconstructing the ANA; Germany was lead for police; the United Kingdom was lead for counternarcotics; Italy was lead for justice; and Japan (with UN assistance) was lead for the disarmament, demobilization, and reintegration of former combatants. In theory, each lead nation was supposed to contribute significant financial assistance, coordinate external assistance, and oversee reconstruction efforts in its sector. In practice, this approach did not work as well as envisioned. The United States provided the bulk of assistance in most security sectors—including counternarcotics, police, and the army.[47] In other areas, such as the justice sector and the disarmament, demobilization, and reintegration of former combatants, there was little measurable improvement.

Counterinsurgency operations are generally complex, demanding, and expensive. Even major powers like the United States need cooperation from others, especially on such issues as basing rights, overflight

---

[46] U.S. Department of Defense, *Information Operations Roadmap* (Washington, D.C.: U.S. Department of Defense, 2003).

[47] Author interviews with U.S., German, and British officials, Kabul, Afghanistan, November 2005 and September 2006.

rights, intelligence, combat forces, economic assistance, and political support.[48] Bilateral donors, international organizations, development banks, and military alliances have different comparative advantages and can provide valuable resources to counterinsurgency. In the absence of broad multilateral support, counterinsurgency operations may not have sufficient military, economic, and political resources to establish security. In addition, the absence of multilateral participation may increase the likelihood that some states will undermine the operation.[49]

NATO's forces in Afghanistan were generally competent. But the NATO experience in Afghanistan highlights several drawbacks with multilateral operations. One was the variation in political will of coalition partners. NATO's International Security Assistance Force was severely limited by the political-military rules of engagement, which constrained each of the national contingents. Some countries, such as Canada and Britain, were reliable allies who were willing to fight—and die—in Afghanistan. During Operation Medusa in 2006, for example, Canadian military forces fought a conventional battle against Taliban forces in Kandahar Province. The Taliban used crew-served weapons and rocket-propelled grenades and engaged from fixed positions.[50] But most NATO countries, such as Germany and Norway, had national caveats that severely restricted their ability to fight. Another drawback was the variation in capabilities. Several coalition countries lacked adequate enabler forces—including attack and lift helicopters, smart munitions, intelligence, engineers, medical, logistics, and digital command and control—to fully leverage and sustain their ground combat power.[51]

---

[48] Richard N. Haass, *Intervention: The Use of American Military Force in the Post–Cold War World* (Washington, D.C.: Carnegie Endowment for International Peace, 2004), pp. 142–147; Robert C. Orr, ed., *Winning the Peace: An American Strategy for Post-Conflict Reconstruction* (Washington, D.C.: The CSIS Press, 2004), pp. 299–301.

[49] Stephen M. Walt, *Taming American Power: The Global Response to U.S. Primacy* (New York: W. W. Norton, 2005), pp. 109–179.

[50] Author interview with Canadian military officials, Kandahar, January 2007.

[51] McCaffrey, "Trip to Afghanistan and Pakistan," p. 4.

A final drawback was a lack of unity of command. In some operations, such as the one carried out in Bosnia following the signing of the 1995 Dayton Peace Accords, the international community created a "high representative" to oversee reconstruction and stabilization. This did not happen in Afghanistan on either the civilian or military side. On the civilian side, there was no unity of command among the international community or U.S. agencies. Even on the military side, there were separate U.S. and NATO chains of command—even after NATO took control over much of the counterinsurgency campaign in 2006. The result was several external forces operating in the same area with different missions and different rules of engagement.

Civil-Military Activities

In his book *Modern Warfare: A French View of Counterinsurgency*, Roger Trinquier argues that counterinsurgency requires

> an interlocking system of actions—political, economic, psychological, military—that aims at the [insurgents' intended] overthrow of the established authority in a country and its replacement by another regime.[52]

One of the most innovative aspects of the Afghan counterinsurgency campaign was its civil-military programs—especially the use of PRTs.[53] PRTs were made up of roughly 60 to 100 soldiers, including civil affairs units, special forces, force protection units, and psychological operations personnel. U.S. PRTs also included less than a half dozen State Department, USAID, and other U.S. government personnel. U.S. PRTs were generally successful in helping strengthen the reach of the central government. They also were successful in providing security at events like the Loya Jirga and during elections. They facilitated recon-

---

[52] Trinquier, *Modern Warfare*, p. 6.

[53] Prior to the establishment of the first PRTs, Coalition Humanitarian Liaison Cells and U.S. Army Civil Affairs Teams–Afghanistan supported humanitarian assistance, relief, and reconstruction efforts throughout Afghanistan. These began in 2002.

struction by funding projects like school repairs or by helping the State Department, USAID, and Department of Agriculture representatives implement civilian-funded projects.[54]

The success of NATO PRTs varied somewhat. They tended to be more successful when working closely with the local population to coordinate reconstruction projects and secure their area of operations. The UK-led PRT in Mazar-e Sharif helped mediate several instances of interfactional fighting and was successful in helping stabilize the region. In contrast, the German-led PRT in Konduz could travel only within a 30-kilometer radius and avoided areas where there was significant factional fighting. Based on interviews with NATO forces, this variation in PRT activity appears at least partly to be a function of domestic politics. In some countries, such as Germany, the aversion to casualties impacted PRT and broader civil-military operations. As one German soldier argued, "The German government is risk averse, especially for any thing more than peacekeeping operations. So is our population. This limits our ability—and willingness—to operate on the ground outside of our bases."[55]

Both U.S. and NATO PRTs faced challenges. One was staffing: Most PRTs were dominated by soldiers, many of whom had little or no development experience. Civilian agencies in all NATO countries had difficulty convincing their personnel to spend time in a war-torn, underdeveloped country. In addition, short tours of duty—including three-month tours—made it difficult for PRT members to gain a full understanding of local politics and culture. There also were too few PRTs. Five years after the overthrow of the Taliban regime, the United States and NATO were able to put PRTs in virtually all major Afghan cities, but they had little operational reach into rural areas.[56] The dete-

---

[54] Robert Borders, "Provincial Reconstruction Teams in Afghanistan: A Model for Post-Conflict Reconstruction and Development," *Journal of Development and Social Transformation*, Vol. 1 (November 2004), pp. 5–12; Michael J. McNerney, "Stabilization and Reconstruction in Afghanistan: Are PRTs a Model or a Muddle?" *Parameters*, Vol. XXXV, No. 4 (Winter 2005–2006), pp. 32–46.

[55] Author interview with German military official, Camp Marmal, Mazar-e-Sharif, September 2006.

[56] McNerney, "Stabilization and Reconstruction in Afghanistan," p. 40.

riorating security environment in the south and east made reconstruction efforts difficult. In early 2007, for example, the Kandahar PRT was involved in reconstruction efforts in several small areas of the province, such as Kandahar City and Panjwari. But it was involved in virtually no reconstruction or development in most of the province because of the security environment. NGOs and state agencies, such as USAID and the Canadian International Development Agency, were also not involved in reconstruction and development in most of the province. The irony in this situation is that rural areas, which were most at risk to the Taliban and where unhappiness with the slow pace of change was greatest among the population, received little or no development assistance.[57]

Team Village missions helped fill in some of this vacuum. The term Team Village refers to a group of personnel—usually a mix of civil affairs and psychological operations personnel—tasked with conducting civil-military operations within a larger campaign. Many also included tactical HUMINT teams, interpreters, military police, media and public affairs personnel, medical personnel, and local Afghan forces.[58] Health care operations were particularly successful in winning support among the locals. U.S. and coalition forces frequently had a line of patients for everything from mild bumps and bruises to more serious injuries and illnesses. Although these patients could be treated from the backs of the high-mobility multipurpose-wheeled vehicles, treatment was easier to administer when given in a secure compound arranged by local elders. Also, it was important to have a female medic available to treat the women and children. Otherwise, women would be either reluctant or forbidden to seek medical help, and they were usually the ones who needed it most.

The use of PRTs, Team Village missions, and other civil-military activities may be one reason that public opinion polls in Afghanistan showed fairly high levels of support for the U.S. and Afghan govern-

---

[57] Author interview with LT COL Simon Heatherington, Commander of the Kandahar Provincial Reconstruction Team, Kandahar, January 16, 2007.

[58] U.S. Army Training and Doctrine Command, *Observations and Lessons Learned: Task Force Devil*, p. 12; Buffaloe, *Conventional Forces in Low-Intensity Conflict*, pp. 13–14.

ment in the early phases of the counterinsurgency campaign.[59] In one poll in 2004, for example, roughly 65 percent of Afghans had a favorable view of the U.S. government, and 67 percent had a favorable view of the U.S. military.[60] In 2005, nearly 70 percent of Afghans rated the work of the United States as either excellent or good. By 2006, nearly 80 percent of Afghans strongly or somewhat supported the work of U.S. military forces.[61] However, there were several indications that support for the Afghan government began to decline in 2006 as levels of violence and frustration with the slow pace of reconstruction increased. As Amrullah Saleh, head of the NDS, concluded, the Taliban succeeded in winning popular support in the east and south and setting up a foothold in such provinces as Helmand and Kandahar. "Who are we?" he asked. "[A] lot of people in the villages of Zabul, Helmand, Kandahar, and Oruzgan have a simple answer to this question. They say this is a corrupt government. They also say you are the government and we are the people. This black and white explanation must change."[62]

Conclusion

In the early stages of the counterinsurgency, U.S and coalition efforts were most effective when they leveraged the Afghan government and indigenous forces, including in such areas as direct action, intelligence, information operations, and civil affairs. This meant operating together with Afghan forces and building Afghan capacity. It was

---

[59] On Afghanistan and public opinion, see The Asia Foundation, *Afghanistan in 2006: A Survey of the Afghan People* (Kabul: The Asia Foundation, 2006); ABC News, *ABC News Poll: Life in Afghanistan* (New York: ABC News, 2005); Morgan Courtney, Frederick Barton, and Bathsehba Crocker, *In the Balance: Measuring Progress in Afghanistan* (Washington, D.C.: Center for Strategic and International Studies, 2005); The Asia Foundation, *Voter Education Planning Survey*; The International Republican Institute, *Afghanistan: Election Day Survey* (Kabul: The International Republican Institute, October 9, 2004).

[60] The Asia Foundation, *Voter Education Planning Survey*, pp. 107–108.

[61] ABC News/BBC World Service, *Afghanistan: Where Things Stand* (Kabul: Afghanistan, 2006), pp. 18–19.

[62] Saleh, *Strategy of Insurgents and Terrorists in Afghanistan*, p. 12.

executed better in some areas than others, and better by some types of forces than others. U.S. special forces tended to be much better trained and prepared to fight in an unconventional environment. There were some areas, such as close air support, in which the United States had to provide most of the capabilities. Developing a robust and technologically advanced Afghan air force that could conduct air strike and air mobility operations proved impractical, in large part because the Afghan Ministry of Defense could not afford it. Chapter Seven will explore in more depth how these findings translate into capabilities for counterinsurgency warfare.

Recommendations

The counterinsurgency campaign in Afghanistan offers a useful opportunity for the U.S. military to develop capabilities to more effectively conduct counterinsurgency operations. The most significant lesson from Afghanistan is the importance of encouraging legitimate and effective indigenous governments and security forces. U.S. military capabilities should focus on leveraging indigenous capabilities and building their capacity to wage a successful counterinsurgency. In some areas, such as air strikes and air mobility, this may be difficult. Most policymakers—including in the United States—repeatedly ignore or underestimate the importance of locals to counterinsurgency warfare. Counterinsurgency requires not only the capability to conduct unconventional war, but also the capability to shape the capacity of the indigenous government and its security forces. The recommendations cover eight areas:

- police
- border security
- ground combat
- air strike and air mobility
- intelligence
- command and control
- information operations
- civil-military activities.

Improvements in all of these areas would increase the capability of indigenous security forces, assist in building governance capacity, and target external support. As noted in Chapter Two, these variables are strongly correlated with the success—and failure—of counterinsurgency efforts. Table 7.1 highlights the eight areas by asking three questions. Who should be the lead actor in each area? What should the U.S. military's role be? What are key U.S. military capabilities necessary to achieve success in this area? While the indigenous government needs to develop an autonomous capacity in all of these areas over the long term, this is not always possible in the near term. It may lack critical capabilities, legitimacy, resources, and political will across key areas—especially in the early phases of a counterinsurgency. Under these conditions, the U.S. military (or another external actor) may need to play a lead role until the indigenous government develops the requisite skills or the insurgency is defeated. This table assumes that the indigenous government will not have sufficient assets in key areas—especially air strike and air mobility.

As Figure 7.1 illustrates, there may also be some variation in how quickly the U.S. military is willing and able to improve indigenous capacity. In some areas, such as policing, there is a strong incentive to build indigenous capacity quickly. There is evidence, for example, that the best and most efficient way to prevent insurgents from gaining ground in the early stages of an insurgency is to have locals do it themselves—what is sometimes called in-group policing.[1] Different communities know their own members, particularly in tightly knit societies in the developing world. These communities thus overcome the intelligence challenge and enable the use of arrests or other forms of pressure with far greater discrimination. In other areas, such as air strike capability, it may not be feasible or desirable to provide the indigenous government with technologically advanced air capabilities.

Police

It should be a key objective of the United States to give primacy to indigenous security forces as much as possible. In some cases, such as

---

[1]   James D. Fearon and David D. Laitin, "Explaining Interethnic Cooperation," *American Political Science Review*, Vol. 90, No. 4 (December 1996), pp. 715–735.

Table 7.1
Example of Counterinsurgency Capabilities

Area	Indigenous Actors	External Actors U.S. Military[a]	External Actors Civilian Agencies[b]	Key U.S. Military Capabilities
Police	Lead	Support	Support	Development of police units to train and mentor indigenous forces, and, when necessary, conduct law enforcement
Border security	Lead	Support	Support	Sensors and other technologies to monitor cross-border activity
Ground combat	Lead (if feasible)	Support	—	Armed reconnaissance capability, with platforms such as AC-130s and tactical unmanned aerial vehicles
Air strike and air mobility	Support	Lead	—	GPS systems and SOFLAMs Satellite radios carried by combat controllers Encrypted radios
Intelligence, surveillance, and reconnaissance	Lead	Support	Support	Ability to quicly train HUMINT assets with language and cultural skill sets and deploy to areas of operation Use of civilian programs, such as medical clinics, for information collection "Gridding" methodology to electronically map location of insurgents, support network

Table 7.1—Continued

Area	External Actors			Key U.S. Military Capabilities
	Indigenous Actors	U.S. Military[a]	Civilian Agencies[b]	
Command and Control	Lead (if feasible)	Support	Support	Empowerment of operations at lowest level
				Train and prepare company and battalion commanders to lead counterinsurgency efforts
				More efficient organizational structures to run external counterinsurgency efforts, such as appointment of civilian administrator
Information operations	Lead	Support	Support	Use of legitimate indigenous groups—such as Muslim clerics and tribal elders—to counter insurgent propaganda
Civil-military activities	Lead	Support	Primary support	Teams to improve corrections system, such as building detention facilities and training
				Ues of PRTs and Team Village units

a In coalition operations, this could also include coalition militaries.

b Civilian agencies could include U.S. agencies, international organizations such as the United Nations, or NGOs.

Figure 7.1
Variation in Indigenous Capacity

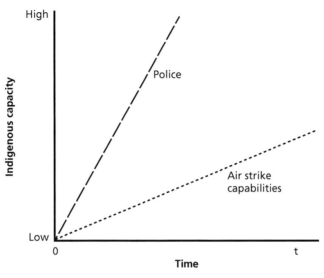

East Timor and Afghanistan, there was virtually no indigenous police capacity when reconstruction and stabilization efforts began. In these cases, the international military or civilian police forces may need to assist in law enforcement missions until this capacity exists and to work as quickly as possible to build a competent indigenous police force. In other cases, such as El Salvador, there was an indigenous police capacity, though it still needed reforms.[2] Over the long run, U.S. forces are unlikely to remain a major combatant for the duration of most counterinsurgencies. The indigenous government and its security forces usu-

---

[2]   On efforts to rebuild the Salvadoran police, see Charles T. Call, "Assessing El Salvador's Transition from Civil War to Peace," in Stephen John Stedman, Donald Rothchild, and Elisabeth Cousens, eds., *Ending Civil Wars: The Implementation of Peace Agreements* (Boulder, Colo.: Lynne Rienner Press, 2002), pp. 383–420; Charles T. Call, "Democratisation, War and State-Building: Constructing the Rule of Law in El Salvador," *Journal of Latin American Studies*, Vol. 35, No. 4 (November 2003), pp. 827–862; William Stanley, "Building New Police Forces in El Salvador and Guatemala: Learning and Counter-Learning," in Tor Tanke Holm and Espen Barth Eide, eds., *Peacebuilding and Police Reform* (Portland, Oreg.: Frank Cass, 2000).

ally know the population and terrain better than U.S. forces do and are more familiar with social and cultural conditions. Also, a lead U.S. role may elicit a strong nationalist or religious backlash among the population. In much of the Muslim world, for example, anti–U.S. sentiment is high. In Egypt and Saudi Arabia, 85 and 89 percent of the populations, respectively, view the United States unfavorably. More than 60 percent of those from Morocco, Jordan, United Arab Emirates, and Lebanon also view the United States unfavorably.[3] This suggests that a large U.S. presence in a predominantly Muslim country may trigger widespread unrest. Finally, a lead indigenous role can provide a focal point for national aspirations in tandem with competent security forces and good governance.

A glaring deficiency in Afghanistan was the poor quality of the police. Due to the nature of their work, police are the first line of defense against insurgents. They have close contact with local populations in cities and villages and will inevitably have a good intelligence picture of insurgent activity. The failure of the police to combat insurgents and undermine their support base can be fatal to counterinsurgency efforts. This means that a major focus of the U.S. military and other government agencies should be to improve the competence of police through training, mentoring, and equipping. There is currently a plethora of U.S. government agencies involved in police training and equipping, such as the Department of Defense, Department of State, Department of Justice, USAID, and CIA.[4] For example, the U.S. Department of State's Bureau of International Narcotics and Law Enforcement and the U.S. Department of Justice's International Criminal Investigative Training Assistance Program have foreign police training programs. However, they are plagued by a paucity of funding and civilian police. This forces the United States to either rely on contractors, such as Dyn-

---

[3]   James Zogby, *2005 Arab Attitudes Toward U.S.: Good News and Bad News* (New York: Zogby International, November 7, 2005).

[4]   On policing during counterinsurgency and stability operations, see Oakley, Dziedzic, and Goldberg, eds., *Policing the New World Disorder*; Jones et al., *Establishing Law and Order After Conflict*; Perito, *Where Is The Lone Ranger When We Need Him?*; Bayley, *Democratizing the Police Abroad*.

Corp, or other states or organizations, such as NATO countries or UN civilian police.

Unlike the police, the ANA's competence improved in the early stages of the counterinsurgency campaign. This was primarily a result of training provided by U.S. and other coalition forces, as well as the integration of indigenous forces into kinetic and nonkinetic operations. The same cannot be said for the ANP, which suffered from a lack of attention, low levels of initial funding, no initial mentoring, corruption, and a paucity of loyalty to the central government. The Department of Defense needs to be involved in rethinking police training. There are at least three options. First, the Department of Defense could create specialized police units within the U.S. military (either the Marine Corps or the Army) or provide predeployment training for an active-duty Army military police brigade. Second, the U.S. government could create police units within a federal law enforcement agency, such as the U.S. Marshals Service, that could be deployed abroad. Third, the U.S. government could fund additional positions within state and selected metropolitan police departments with the understanding that these officers would be available for deployment abroad as part of a federal effort. Additional analysis is necessary to assess these options.[5]

Border Security

Outside assistance was critical to the Afghan insurgents' success in the early stages of the insurgency. Support from the Pakistan government, Pashtun tribes, al Qaeda, and the global jihadist network was crucial to the survivability of insurgent forces. Insurgent groups adapted their tactics and techniques during the counterinsurgency campaign, and became veritable learning organizations. They adopted suicide tactics, increasingly sophisticated IEDs, and networked organizational structures involving small cells.

---

[5] See, for example, Terrence K. Kelly, *Options for Transitional Security Capabilities for America* (Santa Monica, Calif.: RAND Corporation, TR-353-A, 2005).

Since the success of insurgent groups is highly correlated with their ability to acquire external support, the U.S. military needs to improve indigenous capabilities in denying support. Perhaps the most important is improving border security. A major part of the solution is diplomatic and involves negotiations with neighboring countries to curb cross-border activity. In Afghanistan, the United States and other coalition partners failed to alter the will or capacity of the Pakistan government to decrease cross-border activity. Some technological capabilities may improve border security. For uncontrolled border areas, such as the Afghanistan-Pakistan border, these capabilities might include a range of sensors and platforms to detect insurgent movement. Efforts along these lines might also include training and equipping indigenous forces to conduct patrols, utilize a range of technologies, interdict insurgent forces, and work with neighboring governments to enforce border security.

There are two related areas in which capabilities are needed to deny sanctuary and target external support. One is countering insurgent recruitment. The military and political authorities need to utilize diplomatic and other means to encourage neighboring states to curb recruitment campaigns for insurgents, close training camps, and conduct a sustained counterinsurgency campaign that undermines popular support for insurgents and captures or kills leaders and guerrillas. Insurgents' ability to maintain sanctuary in neighboring countries can be lethal to a successful counterinsurgency effort. In Afghanistan, counterinsurgency efforts were limited largely to Afghan territory and did little to curb recruitment and other support in Pakistan and other neighboring countries.

The second area requiring new capabilities is in countering the external financing of insurgent groups. The U.S. government needs to continue pressuring foreign governments to curb the financing of insurgent groups from diaspora populations and wealthy financiers in such regions as the Persian Gulf—in close cooperation with the Department of State, Department of Treasury, and other U.S. and coalition partners. Afghan groups received a steady stream of assistance from wealthy individuals in the Gulf and other regions. This funding stream was an important resource for insurgent groups.

Ground Combat

Military force is usually most effective when used in ways that are considered legitimate; it is also more effective when used by locals and not external actors. The most effective use of military force in a counterinsurgency includes two types of ground combat capabilities.

First, one of the most successful approaches in Afghanistan— which was used to varying degrees—was clear, hold, and expand. This is sometimes referred to as an ink-spot strategy. Forces are assigned to contested areas to regain government presence and control and then conduct military and civil-military programs to expand the control and edge out insurgents. The focus is on consolidating and holding ground that is clearly progovernment; protecting the government and other key resources (such as lines of communication and major cities); and deploying counterinsurgency forces to conduct offensive operations in contested areas. Holding territory proved to be the most difficult facet of clear, hold, and expand in Afghanistan due to the low number of U.S. and local forces.[6] This suggests that sufficient numbers of forces are needed to hold territory once it is cleared, or insurgents can retake it. Deploying forces into insurgent areas can deny them sanctuary, interdict the border, and expand government and coalition presence. Where possible, U.S. counterinsurgency forces should be kept to a minimum and supported with civil-affairs and psychological operations units. A company of infantry can sometimes be provided for area patrolling and security for an immediate threat to the unit. Quick reaction forces in the form of close air support assets or reinforcing units should back up the outposts whenever insurgent forces threatened to overrun them.

Clear, hold, and expand involves conducting operations in ever-increasing zones around military bases. The first measure for force protection is to target and eliminate the insurgents living within the inner zone—defined by mission, enemy, troops, terrain, and time around their base. This requires living among the local population for long

---

[6]   On the low number of U.S. and coalition forces, see Jones, "Averting Failure in Afghanistan"; James Dobbins, Seth G. Jones, Keith Crane, Andrew Rathmell, Brett Steele, Richard Teltschik, and Anga Timilsina, *America's Role in Nation Building: From Germany to Iraq* (Santa Monica, Calif.: RAND Corporation, MR-1753-RC, 2003).

durations to gain trust and support and the ability to separate the locals from the insurgents. The secondary zone is the transit and support zone for the insurgents. It may include remote locations or areas where the population is neither friendly nor hostile to the counterinsurgency unit's efforts. Occasional operations need to be conducted in these areas to show the flag and keep the population neutral to the idea of supporting the insurgents. In Afghanistan, battalion-sized sweeps and clearing operations by conventional forces generally reaped far less than their effort warranted because of the difficulty of finding and fixing elusive insurgents.[7]

Second is the development of an armed reconnaissance capability and a specialized raiding force. Armed reconnaissance is the patrolling of suspected insurgent areas to glean information on their activities, initiate contact, or confirm the area is clear. Armed reconnaissance can be accomplished with a variety of platforms and measures, such as AC-130 gunships (operating generally at night), tactical unmanned aerial vehicles, and mounted ground reconnaissance patrols.[8] A specialized raiding force is sometimes required to conduct time-sensitive targeting beyond the scope of conventional forces. These specialized raiding forces may consist of a counterterrorist unit, an indigenous strike force, or a specially formed and trained unit drawn from organic forces. In Afghanistan, raiding forces often required dedicated mobility platforms and a high level of access to intelligence assets. The sensor-to-shooter links often worked best when noncontributing layers of decisionmakers were removed.

Air Strike and Air Mobility

Air strike and air mobility are also important. The effectiveness of close air support in counterinsurgency operations such as those conducted in Afghanistan made it particularly useful for the United States to continue to develop technological capabilities such as GPS-based systems

---

[7]   Celeski, *Operationalizing COIN.*

[8]   Burda, *Operation Enduring Freedom Lessons Learned.*

and SOFLAMs, which were invaluable to ground forces. For much the same reason, it was also useful to continue to develop new and more sophisticated communication equipment, including advanced receivers and transmitters that link forces in the field to those in other areas, satellite radios carried by combat controllers to call in air strikes, and encrypted HF radio.[9] Air mobility capabilities are also likely to be important. In most developing countries, indigenous forces are unlikely to possess significant air mobility assets, and the United States (or other external actors) may need to provide air transport to move forces into (and out of) the area of operations.

Close air support provided a particularly significant advantage to small groups of U.S. and Afghan forces operating against insurgents. This lesson may not be applicable to all counterinsurgency operations, especially those conducted in urban areas. In Afghanistan, most of the fighting occurred in rural areas in which close air support was possible and effective. A variety of aircraft can provide close air support to U.S. and indigenous forces, including AH-64 attack helicopters, Spectre AC-130 gunships, A-10 and F-14 fighters, and B-52 bombers.[10]

Intelligence

The problem of destroying insurgent groups and their supporters is often one of finding them. This makes access to reliable and action-able information of paramount importance. The U.S. military needs to continue developing capabilities for collecting, analyzing, and acting quickly on intelligence. In Afghanistan, the U.S. experience suggests that there are limits to applying technological solutions to intelligence challenges. Advanced technologies were sometimes helpful in identify-ing insurgents. However, insurgents were frequently able to adapt their tactics, techniques, and procedures to avoid these collection methods. This makes HUMINT particularly important—including HUMINT

---

[9] On the utility of SOFLAMs and communication equipment in Afghanistan, see Berntsen and Pezzullo, *Jawbreaker*, pp. 78–79, 83, 134, 266–268.

[10] Pirnie et al., *Beyond Close Air Support*.

gathered by indigenous intelligence and security services, as well as local tribes and militias. Civil-military programs—such as those providing medical assistance to locals—can also be utilized to build trust with locals. The development of analytical tools to process information is also important. For example, the utilization of a gridding system to map the location of insurgents and their support network has been helpful in processing intelligence into a useful operational planning tool for the U.S. military in Afghanistan.

Effective analysis capability is a critical component of any intelligence capability. Counterinsurgency operations require the development of an analytical methodology to measure the insurgency's impact on the local population—especially the impact of the security condition. A number of factors can make it difficult to measure the effectiveness of counterinsurgency operations: Progress cannot be measured by the advance of militaries across a map as in conventional warfare; focusing only on guerrilla fighters misses the much broader support network; a complicated array of political, economic, social, and military factors can fuel the insurgency; and there are rarely ideal predefined qualitative or quantitative target metrics.

Unfortunately, the counterinsurgency campaign in Afghanistan often fell into two measurements traps: (1) measuring success based on outputs and inputs rather than on broader strategic outcomes; and (2) focusing on U.S., rather than Afghan, metrics. The U.S. military needs to develop outcome-based metrics to assess performance during counterinsurgency campaigns. Inputs should refer to the amount of resources used in counterinsurgency operations, such as the amount of financial assistance and soldiers deployed. Outputs should be the first-order results of the counterinsurgency program. These include such metrics as indigenous forces trained and insurgents killed or captured. Outcomes are conditions that directly impact the local population. They should not measure what the military does, but rather represent the consequences of its efforts.[11] Without such an ability to measure performance, policymakers lack an objective method for judg-

---

[11] Jones et al., *Establishing Law and Order After Conflict*.

ing success and failure in ongoing crises, making midcourse correc-
tions more difficult.

Despite these hurdles, there are reasonable measures of counterinsurgency effectiveness. Key outcome indicators include the number
and lethality of attacks against the local population over time and
trends in public opinion polls. Metrics need to be tied to the security of the population. The tools to assess counterinsurgency outcomes
remain limited. It is difficult to measure popular support and security,
especially in countries such as Afghanistan that have not collected sys-
tematic data. However, by building such assessments into current and
future assistance programs and encouraging host nations to undertake
such assessments, the U.S. military will be better placed to optimize
counterinsurgency operations.[12] Based on the Afghanistan case, key
outcome indicators might include tracking the following:

- *Insurgent violence.* The most useful indicators here are trends over
  time in the number and lethality of insurgent-initiated attacks
  against the population. First, how many attacks have insurgents
  conducted and where have the conducted them? Are they rising
  or declining over time, and is their area of operations spreading
  or shrinking? Second, how lethal are their attacks over time? Are
  they able to kill more people (especially locals)? Short-term trends
  are generally not useful, since the rate of enemy attacks can fluc-
  tuate and a decline in the rate may be caused by preparation for a
  larger offensive.[13] Crime data, including homicide rates over time,
  are also useful.
- *Public opinion.* Useful indicators in this area include trends in
  public opinion polls that capture the perception of security among
  locals. Does the local population feel more (or less) safe over time?
  Why? Public perceptions of security can be an important indica-

---

[12] For instance, public opinion polling can be adjusted to address perceptions of corruption or judicial integrity. International advisors can assist in the development of system-wide
criminal justice metrics systems as they undertake routine projects.

[13] Richard Betts, *Stability in Iraq?* (Washington, D.C.: Central Intelligence Agency, 2005),
p. 4.

tor of the effectiveness of the security forces in establishing an environment in which people and goods can circulate freely and licit political and economic activity can occur without intimidation. Promoting disorder is a key objective for most insurgents. Disrupting the economy and decreasing security help produce discontent among the government and undermine the strength and legitimacy of the indigenous government.[14] Evidence of a relaxation of behavior stemming from fear can be particularly instructive: Do locals go out at night in their villages? Do they travel outside their villages at night?

- *Other indicators of security.* More tactical outcome measures may also be appropriate. In Somalia, for example, the U.S. military collected such indicators as the death rate per day due to starvation, new patients with gunshot wounds in hospitals, and the street price of an AK-47.[15] All outcome measures, however, should be tied to the overall mission objectives.

The ability to electronically map territory to monitor insurgent sanctuaries and their support network is another critical intelligence capability.[16] In Afghanistan, a gridding system was used fairly successfully to cover large sectors of suspected insurgent territory, as illustrated in Figure 7.2. Each square in the grid was covered by one method of surveillance and either confirmed as empty or contained suspicious activity. Once suspicious activity was noted, U.S. or coalition forces continued to develop the situation, sometimes introducing additional forces in the area to destroy the threat. Timing was critical. Long hours and numerous assets were required to find insurgent forces. Hunter-killer platforms engaged insurgent forces immediately upon their discovery and then continued to maintain contact while reinforcements were sent to the area.

---

[14] Galula, *Counterinsurgency Warfare*, pp. 11–12, 78–79; Jones et al., *Establishing Law and Order After Conflict.*

[15] U.S. Marine Corps, *Small Wars Manual.*

[16] Trinquier, *Modern Warfare*; Robert R. Tomes, "Relearning Counterinsurgency Warfare," *Parameters*, Vol. XXXIV, No. 1 (Spring, 2004), pp. 16–28.

Figure 7.2
Example of Grid Methodology

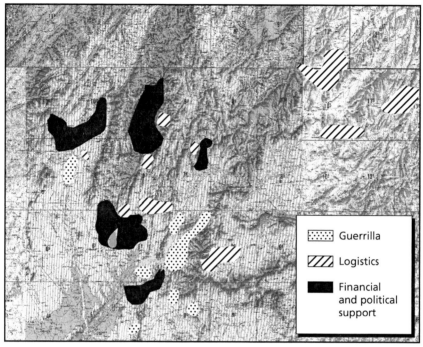

Guerrilla

Logistics

Financial
and political
support

RAND *MG595-7.2*

The gridding methodology is roughly comparable to the intelligence tactics used by the British in Malaya and United States in Vietnam. In Vietnam, Secretary of Defense Robert Strange McNamara asked the CIA in 1966 to develop a technique to measure trends in "pacification" of inhabitants. The result was the Hamlet Evaluation System, which contained such objectives as whether government or insurgent forces were in a hamlet. However, it relied heavily on subjective measures to gauge the quality of life of the people. The Hamlet Evaluation System reported whether hamlet-dwellers lived in some tranquility or if they were subject to harassment, intimidation, and attacks from the Viet Cong or North Vietnamese Army.

The most important feature of the Hamlet Evaluation System was the fact that two separate parties prepared two separate reports independently of one another. The U.S. Senior District Advisor and his counterpart, the Vietnamese District Commander, each prepared a report for his respective chain of command. This was intended to ensure that the two did not cooperate to bias the conclusions. After the reports were submitted, they were compared at the Province Senior Advisor level, then at the Military Assistance Command–Vietnam/ Civil Operations for Revolutionary Development Support headquarters in Saigon. The questions were given a mathematical compilation to provide a statistical average for each hamlet. These were then accumulated for each village, city, district, and province. A hamlet or village that was rated secure or "under government influence" was classified as A, B, or C. Those hamlets under guerrilla influence were given a D or E. Analyzing data from the Hamlet Evaluation System, as well as data on unit operations, showed how maneuver battalion days of operations in a given area contributed little to population security. Big-unit sweeps did not promote pacification unless U.S. forces stayed in the area.[17]

Command and Control

The mission profiles of future counterinsurgency operations may require adapting organizational structures in several ways. One is to empower operations at the lowest level. Well-trained small-unit maneuver is important to success.[18] Competent insurgent groups can disperse their forces, make them smaller and more difficult to attack, and acquire more secure communication, better camouflage, and more effective diversions.[19] U.S. military operations may be more likely to succeed when leaders at the small-unit level have enough leeway, specialized

---

[17] Andrew F. Krepinevich, *The Army and Vietnam* (Baltimore, Md.: Johns Hopkins University Press, 1986).

[18] U.S. Army Training and Doctrine Command, *Operation Enduring Freedom*, pp. 22–23.

[19] Biddle, *Afghanistan and the Future of Warfare*; Cordesman, *The Ongoing Lessons of Afghanistan*, pp. 122–123.

assets, and firepower to engage the population, develop their own intelligence, and defeat the enemy in combat.[20]

Indeed, U.S. military doctrine needs to establish far looser and more broadly distributed networks that have a high degree of individual independence and survivability.[21] This means incorporating into counterinsurgency doctrine and training the preparation of company and battalion commanders to lead combined-arms warfare, conduct civil-military operations, and develop and exploit their own intelligence. It also means giving infantry commanders the responsibility, autonomy, and distance from higher headquarters that is now held only by special forces A-team commanders. Commanders must empower small-unit leaders to deal with the challenges encountered during counterinsurgency operations, including the authority to routinely make decisions currently made by battalion and brigade combat team commanders.[22] Command and control works well when it has been flattened out from a hierarchical to a more horizontal level. The shorter sensor-to-shooter links are, the better they work. Quicker and more responsive arrangements for command and control provide flexibility for forces on the battlefield.[23] This was not always well executed in Afghanistan.

At the strategic level, command and control presented three significant challenges in Afghanistan. One was overall coordination of the counterinsurgency. Where possible, the indigenous government should be in charge of counterinsurgency efforts overall. While international civilian personnel and military forces will retain separate chains of command from the indigenous government, international actors need to support the preeminent role of the indigenous government.

The second strategic-level challenge concerns command and control among international forces. As noted Chapter Six, there were separate U.S. and NATO chains of command. The result was several external forces operating in the same area with different missions and

---

[20] Buffaloe, *Conventional Forces in Low-Intensity Conflict*, p. 4.

[21] Cordesman, *The Ongoing Lessons of Afghanistan*.

[22] U.S. Army Training and Doctrine Command, *Observations and Lessons Learned: Task Force Devil*, p. 2.

[23] Celeski, *Operationalizing COIN*, p. 83.

different rules of engagement. As demonstrated in other cases, such as Somalia and Sierra Leone, this practice engenders duplication and risks confusion, miscommunication, and, in some instances, unnecessary casualties. A better model may be that followed by the United States in Haiti in 1994 and Australia in East Timor in 1999, in which the U.S.- and Australian-led entry forces, respectively, were drawn down and incorporated into a unified force with a single chain of command.[24]

The third strategic-level command and control challenge in the Afghan counterinsurgency was a lack of coordination within the U.S. effort. Successful counterinsurgency campaigns require a combination of military and political capabilities. But coordination between the U.S. military and civilian agencies operating in Afghanistan was problematic. This suggests a need to get civilian agencies better engaged and coordinated on the ground with the military. Overcoming this challenge will also require rethinking the organizational structure of the counterinsurgency campaign. There are several potential options, though additional analysis is necessary to assess them. The first is to put a civilian in charge of the counterinsurgency effort, perhaps the U.S. ambassador in the country in which the insurgency is taking place. Or, following the model of several British-led counterinsurgencies, it could be a special envoy (or high commissioner) with command and control authority. Under both of these options, the top military official would be subordinate to the civilian figure. Indeed, British civilian officials were in control of emergencies and were responsible for the broader political strategy. The British Army operated under civilian control and accepted the requirement of employing minimum force. Another option is to give the military command and control responsibilities for the counterinsurgency campaign. The final option is to establish an interagency working group, which includes both civilian and military officials, that has overall responsibility for the counterinsurgency campaign.

---

[24] See, for example, James Dobbins, Seth G. Jones, Keith Crane, and Beth Cole DeGrasse, *The Beginner's Guide to Nation-Building* (Santa Monica, Calif.: RAND Corporation, MG-557-SRF, 2007).

Information Operations

Indigenous groups are more likely to be effective in influencing locals and countering insurgent ideology than the U.S. military or other international actors are. It is critical to understand who holds power, who the local population trusts, and where locals get their information—and then to target these forums. In some cases, such as in Afghanistan, religious leaders and tribal elders wield the vast majority of power. This means providing assistance to credible indigenous groups, such as Muslim clerics or tribal elders, that can effectively counter jihadist propaganda. These groups do not necessarily have to be supportive of the United States, but they do need to oppose insurgents and have credible influence among the population. Much of this funding may have to be indirect to protect their credibility. Assistance could be directed to indigenous media, political parties, student and youth organizations, labor unions, and religious figures and organizations that meet at least two criteria: (1) they have a notable support base in the local population; and (2) they oppose insurgent groups and insurgent ideology. This approach has some parallels with U.S. efforts during the Cold War to balance the Soviet Union by funding existing political, cultural, social, and media organizations in areas such as Central and Eastern Europe.

Insurgents sometimes make religion a significant part of their rhetoric to gain popular support. Since mosques have historically served as a tipping point for major political upheavals in Afghanistan, Afghan government officials have targeted mosques. The Ulema Council of Afghanistan called on the Taliban to abandon violence and support the Afghan government in the name of Islam. They also called on the religious scholars of neighboring countries—including Pakistan—to help counter the activities and ideology of the Taliban and other insurgent organizations. A number of Afghan Islamic clerics publicly supported the Afghan government and called the jihad un-Islamic. Moreover, the Ulema Council and a number of Afghan ulema have issued fatwas, or religious decrees, that unambiguously oppose suicide bombing. Indigenous efforts to influence the population, especially credible actors such as Muslim clerics and tribal leaders, were the most successful.

Civil-Military Activities

The Afghan government suffered from significant governance problems, such as its failure to stem the rising drug trade and corruption
in key areas like the justice system. While civilian agencies should
lead efforts to rebuild the justice system, the military needs to play an
important role as well. This means establishing civil-affairs teams that
can work in key areas of the justice system, such as establishing courts
to prosecute insurgents, terrorists, and major criminals involved in
such activities as drug trafficking. Effectively prosecuting "spoilers"—
including insurgents—is critical to the success of a counterinsurgency
campaign. The failure to adequately deal with these individuals undermines counterinsurgency efforts. In addition, pretrial detention facilities and prisons are often in poor condition in developing countries.
Compounding this problem, donor states may be reluctant to become
involved in rebuilding a system with a reputation for significant human
rights violations. The military can help fill this gap.

Reconstruction teams should be a core component of counterinsurgency operations, and involve both military and civilian personnel.[25]
In the U.S. government, this includes personnel from such organizations as USAID and the Departments of State, Justice, Treasury, and
Agriculture. These teams should become involved in a wide variety of
projects that range from assisting the local population with basic health
care; helping improve basic services such as water and electricity; and
rebuilding schools, government buildings, and hospitals. Reconstruction programs should be designed to gain support for the indigenous
government, rather than for the United States or other external actors.
As Israeli General Moshe Dayan argued, "foreign troops never win the
hearts of the people" and generally do not gain support for the host
government.[26] In areas where there is significant insurgent violence,
the U.S. military may need to provide a greater percentage of civil-

---

[25] Khalilzad, "How to Nation-Build: Ten Lessons From Afghanistan."

[26] Quoted in Wilensky, *Military Medicine to Win Hearts and Minds*, p. 132. Also see General
Moshe Dayan, Folder 488, 1-31 May 1968, #32 History File, Box 34, Record Group 319,
National Archives and Records Administration.

military assistance, since they are better prepared to operate in hostile environments.

These nonkinetic operations are a critical weapon in an indigenous government's battle for popular support. The U.S. military should deploy civil affairs units with training and experience in such areas as public administration, public safety, public health, and legal systems. Again, the focus of the U.S. military and other international actors should be to improve the indigenous government's ability to rebuild and provide essential services—not to do it for them. This is a key aspect of any attempt to secure popular support and legitimacy for the government. One of the most innovative aspects of the Afghan counterinsurgency campaign was the inclusion of civil-military programs—especially the use of PRTs. The PRT concept can help strengthen the reach of the central government and rebuild essential services. PRTs can facilitate reconstruction by funding projects such as school repairs or by helping Department of State, USAID, and Department of Agriculture representatives to implement civilian-funded projects. However, there are several barriers to the successful implementation of PRTs. One is staffing hurdles. Short tours of duty—such as three-month tours—make it difficult for PRT members to develop an understanding of local politics and culture. A second is a scarcity of PRTs, which need to reach beyond urban areas into rural areas, where insurgent support may be strongest. In Afghanistan, they had little operational reach in rural areas. Since insurgents are often better able to build support networks and capture territory in rural areas, PRTs need to spread to rural areas.

This is where Team Village missions can be effective. In Afghanistan, Team Village units—which usually included a mix of civil affairs and psychological operations personnel—were tasked with conducting civil-military operations within a larger campaign. Team Villages also included tactical HUMINT teams, interpreters, military police, media and public affairs personnel, medical personnel, and local Afghan forces.[27] Health care operations can be particularly successful in winning

[27] U.S. Army Training and Doctrine Command, *Observations and Lessons Learned: Task Force Devil*, p. 12; Buffaloe, *Conventional Forces in Low-Intensity Conflict*, pp. 13–14.

support among the locals. Local villagers can be treated for everything from mild bumps and bruises to more serious injuries and illnesses. Although these patients can be treated from the backs of the high-mobility multipurpose-wheeled vehicles, such care is better tended in a secure compound arranged by the locals. Also, it is key to have a female medic available to treat the women and children.

Good governance involves the provision of essential services to the population by a central authority in a timely manner. This includes the process by which those in authority are selected, monitored, and replaced (the political dimension); the government's capacity to effectively manage its economic resources and implement sound policies (the economic dimension); and the respect of citizens and the state for the country's institutions (the institutional dimension).[28] The U.S. military is generally not the primary actor for governance and institution-building during counterinsurgency operations. Nonetheless, it should play a critical role in encouraging good governance by helping the indigenous authorities deal with corruption, improve the criminal justice system, and disarm militias. In the long term, poor governance can undermine a counterinsurgency effort by undermining popular support for the government. Thus it is important to ensure coordination between other actors involved in governance: UN agencies such as the UN Development Programme; regional organizations such as the Asian Development Bank, the European Union, and the Organization for Security and Cooperation in Europe; and a broad range of non-governmental organizations. The failure of governance in key provinces—such as Helmand and Kandahar—has seriously undermined counterinsurgency efforts in Afghanistan.

In 1940, the U.S. Marine Corps' Small Wars Manual perceptively noted that "the history of the United States shows that in spite of the varying trend of the foreign policy of succeeding administrations, this Government has interposed or intervened in the affairs of other states with remarkable regularity." It continued with the contention that since these types of missions would likely recur in the future, "it is well that the United States may be prepared for any emergency

---

[28] Kaufmann, "Myths and Realities of Governance and Corruption."

which may occur."[29] This assessment is just as true today as it was in the 1940s. Indeed, the United States has been involved in numerous counterinsurgencies throughout its history in Latin America, the Middle East, Africa, Europe, and Asia. The challenge is to improve its ability to engage more effectively in counterinsurgency operations in the future.

[29] U.S. Marine Corps, *Small Wars Manual*.

Insurgencies Since 1945

China	1946–1950
Greece	1945–1949
Philippines	1946–1952
Vietnam	1945–1954
Burma	1948–
Malaya	1950–1956
Colombia	1948–1962
China/Tibet	1950–1951
Kenya (Mau Mau)	1952–1956
Cuba	1958–1959
Algeria	1954–1962
Lebanon	1958–1958
Indonesia (Darul Islam)	1958–1960
Congo/Katanga	1960–1965
Guatemala	1968–1996
South Africa (African National Congress)	1983–1994
Ethiopia/Eritrea	1974–1992
Laos	1960–1973

Vietnam	1960–1975
Iraq (Kurds)	1961–1974
Yemen	1962–1969
Mozambique	1964–1974
Guinea–Bissau	1962–1974
Angola	1962–1974
Colombia	1963–
Zimbabwe (Rhodesia)	1972–1979
Dominican Rep.	1965–1965
Nigeria/Biafra	1967–1970
Argentina (Montañeros)	1973–1977
Cambodia	1970–1975
Northern Ireland	1969–1999
Philippines (New Peoples Army)	1972–1994
Jordan (Black September)	1970–1970
Philippines (Moro National Liberation Front)	1968–
Pakistan/Bangladesh	1971–1971
Pakistan/Baluchistan	1973–1977
Angola (Unita)	1975–
Morocco (Polisario)	1975–1988
Indonesia/East Timor	1975–1999
Namibia	1960–1991
Uruguay	1963–1973
Philippines (Moro Islamic Liberation Front)	1977–

India Naxalite	1980–
Nigeria/Biafra II	1999–
Afghanistan	1996–2001
Lebanon	1975–1990
India Northeast	1952–
Indonesia/Aceh	1999–
Mozambique (Renamo)	1976–1995
Sri Lanka (Liberation Tigers of Tamil Eelam)	1983–
Nicaragua	1978–1979
Afghanistan	1978–1992
Cambodia	1978–1992
El Salvador	1979–1992
Somalia (anti-Barre)	1981–1991
Senegal	1989–
Peru	1981–1992
Nicaragua (Contras)	1981–1988
Turkey (*Partiya Karkerên Kurdistan*)	1984–1999
Sudan (Sudan People's Liberation Army)	1983–
Uganda	1993–
Papua New Guinea/Bougainville	1988–1998
Liberia	1989–1996
India/Kashmir	1989–
China/Xinjiang	1991–
Rwanda	1990–

Moldova	1992–1994
Sierra Leone	1991–
Somalia	1991–
Algeria (*Groupe Islamique Armé*)	1992–
Croatia/Krajina	1992–1995
Afghanistan	1992–1901
Tajikistan	1992–1997
Georgia/Abkhazia	1992–1994
Azerbaijan/Ngo-Kar	1992–1994
Bosnia	1992–1995
Burundi	1993–
Pakistan (Sindhis versus Mahajirs)	1993–1999
Chechnya	1994–1996
Congo (Kabila)	1996–1997
Nepal	1997–
Congo	1998–
Chechnya II	1999–
Serbia/Kosovo	1998–1999
Nigeria/communal	1999–
Israel	1996–
Afghanistan	2001–
Ivory Coast	2002–
Sudan/Darfur	2002–
Iraq	2003–

References

ABC News. *ABC News Poll: Life in Afghanistan*. New York: ABC News, 2005.

ABC News/BBC World Service. *Afghanistan: Where Things Stand*. Kabul, Afghanistan, 2006, pp. 18–19.

Afghan Islamic Press, Interview with Mofti Latifollah Hakimi, August 30, 2005.

Afghan Non-Governmental Organization Security Office. *Security Incident—Armed Clash: ANP Was Disarmed*. Kabul: Afghan Non-Governmental Organization Security Office, March 2006.

Afghan Spokesman Calls on Pakistan to Curb Taliban Activities. Kabul Tolu Television, June 21, 2005.

Afghan Taliban Report Execution of Two People on Charges of Spying for U.S. Afghan Islamic Press, July 12, 2005.

Afghan Taliban Say No Talks Held with U.S., No Differences with Hekmatyar. *Karachi Islam*, February 24, 2005: pp. 1, 6.

Afghanistan in 1986: The Balance Endures. *Asian Survey*, Vol. 27, No. 2: pp. 127–142.

Afghanistan Ministry of Defense. *Securing Afghanistan's Future: Accomplishments and the Strategic Path Forward, National Army*. Kabul: Ministry of Defense, 2004.

———. *The National Military Strategy*. Kabul: Afghanistan Ministry of Defense, October 2005.

Afghanistan Ministry of Interior. *Afghan National Police Program*. Kabul: Ministry of Interior, 2005.

Afghanistan National Security Council. *National Threat Assessment*. Kabul: National Security Council, 2005.

Alexiev, Alex. Tablighi Jamaat: Jihad's Stealthy Legions. *Middle East Quarterly*, Vol. XII, No. 1 (Winter 2005).

Al Jazeera Airs Hikmatyar Video. Al Jazeera TV, May 4, 2006.

Al Qaeda in Afghanistan. The Rule of Allah. Video, 2006.

al-Zawahiri, Ayman. *Knights Under the Prophets Banner.* n.p., December 2001.

Amir, Intikhab. Waziristan: No Man's Land. *The Herald* (Pakistan), April 2006: p. 78.

———. Whose Writ Is It Anyway? *The Herald* (Pakistan), April 2006: pp. 80–82.

Anderson, John Ward. Arms Seized in Afghanistan Sent From Iran, NATO Says. *Washington Post*, September 21, 2007, p. A12.

The Asia Foundation. *Voter Education Planning Survey: Afghanistan 2004 National Elections.* Kabul: The Asia Foundation, 2004.

———. *Afghanistan in 2006: A Survey of the Afghan People.* Kabul: The Asia Foundation, 2006.

Art, Robert J. "Europe Hedges Its Security Bets." In *Balance of Power Revisited: Theory and Practice in the 21st Century*, edited by T. V. Paul, James Wirtz, and Michel Fortmann, eds. Palo Alto, Calif.: Stanford Universitty Press, 2004.

Asian Development Bank and World Bank. *Afghanistan: Preliminary Needs Assessment for Recovery and Reconstruction*, Kabul: Asian Development Bank and World Bank, January 2002.

Barno, LTG David W. *Afghanistan: The Security Outlook.* Washington, D.C.: Center for Strategic and International Studies, May 14, 2004.

Bayley, David H. *Democratizing the Police Abroad: What to Do and How to Do It.* National Institute of Justice, June 2001.

Beblawi, Hazem, and Giacomo Luciani, eds. *The Rentier State.* New York: Croom Helm, 1987.

Berntsen, Gary, and Ralph Pezzullo. *Jawbreaker: The Attack on Bin Laden and Al Qaeda.* New York: Crown Publishers, 2005.

Betts, Richard K. *Stability in Iraq?* Washington, D.C.: Central Intelligence Agency, 2005.

Bhatia, Michael, Kevin Lanigan, and Philip Wilkinson. *Minimal Investments, Minimal Results: The Failure of Security Policy in Afghanistan.* Kabul: Afghanistan Research and Evaluation Unit, June 2004.

Biddle, Stephen. *Afghanistan and the Future of Warfare: Implications for Army and Defense Policy.* Carlisle, Pa.: Strategic Studies Institute, U.S. Army War College, November 2002.

Bloom, Mia. *Dying to Kill: The Allure of Suicide Terror.* New York: Columbia University Press, 2005.

de Borchgrave, Arnaud. Talibanization of Pakistan. *Washington Times*, April 7, 2007, p. A11.

Border Roads Organisation. *Vision, Mission, Role*. Delhi: Border Roads Organisation, 2006.

Borders, Robert. Provincial Reconstruction Teams in Afghanistan: A Model for Post-Conflict Reconstruction and Development. *Journal of Development and Social Transformation*, Vol. 1 (November 2004).

Buffaloe, David L. *Conventional Forces in Low-Intensity Conflict: The 82d Airborne in Firebase Shkin*. Landpower Essay 04-2. Arlington, Va.: Association of the United States Army, 2004.

Burda, COL Bruce. *Operation Enduring Freedom Lessons Learned*. Hurlburt Field, Fla.: Air Force Special Operations Command, 2003.

Burke, Jason. The New Taliban, *Observer* (London), October 14, 2007.

Byman, Daniel L. *Deadly Connections: States That Sponsor Terrorism*. New York: Cambridge University Press, 2005.

——. *Going to War with the Allies You Have: Allies, Counterinsurgency, and the War on Terrorism*. Carlisle, Pa.: U.S. Army War College, November 2005.

——. Friends Like These: Counterinsurgency and the War on Terrorism. *International Security*, Vol. 31, No. 2 (Fall 2006): 79–115 .

Byman, Daniel L., Peter Chalk, Bruce Hoffman, William Rosenau, and David Brannan. *Trends in Outside Support for Insurgent Movements*. Santa Monica, Calif.: RAND Corporation, MR-1405-OTI, 2001. As of January 19, 2007: http://www.rand.org/pubs/monograph_reports/MR1405/

Byrd, William, and Christopher Ward. *Drugs and Development in Afghanistan*. Washington, D.C.: World Bank, 2004.

Call, Charles T. Assessing El Salvador's Transition from Civil War to Peace. In Stephen John Stedman, Donald Rothchild, and Elisabeth Cousens, eds., *Ending Civil Wars: The Implementation of Peace Agreements*. Boulder, Colo.: Lynne Rienner Press, 2002.

——. Democratisation, War and State-Building: Constructing the Rule of Law in El Salvador. *Journal of Latin American Studies*, Vol. 35, No. 4 (November 2003): pp. 827–862.

Callwell, Colonel C. E. *Small Wars: Their Principles and Practice*, 3rd ed. Lincoln, Neb.: University of Nebraska Press, 1996.

Canadian Soldier Dies in Suicide Attack in Kandahar. Afghan Islamic Press, March 3, 2006.

Celeski, Joseph D. *Operationalizing COIN*. JSOU Report 05-2. Hurlburt Field, Fla.: Joint Special Operations University, 2005.

Center for Army Lessons Learned. *Ranger Observations from OEF and OIF: Tactics, Techniques, and Procedures*. Fort Leavenworth, Kan.: Center for Army Lessons Learned, February 2005.

Central Intelligence Agency. *Guide to the Analysis of Insurgency*. Washington, D.C.: Central Intelligence Agency, n.d.

Clarke, Richard. *Against All Enemies: Inside America's War on Terror*. New York: Free Press, 2004.

Coalition Launches "Operation Mountain Lion" in Afghanistan. American Forces Press Service, April 12, 2006.

Coll, Steve. *Ghost Wars: The Secret History of the CIA, Afghanistan, and bin Laden, from the Soviet Invasion to September 10, 2001*. New York: Penguin Books, 2004.

Collier, David. The Comparative Method: Two Decades of Change. In *Comparative Political Dynamics: Global Research Perspectives*, edited by Dankwart A. Rustow and Kenneth Paul Erickson. New York: Harper Collins, 1991.

Collier, David and James Mahoney. Insights and Pitfalls: Selection Bias in Qualitative Research. *World Politics*, Vol. 49, No. 1 (October 1996): pp. 56–91.

Combined Forces Command—Afghanistan. *Afghan National Security Forces Operational Primacy Process*. Kabul, Afghanistan: Combined Forces Command—Afghanistan, 2006.

Combined Joint Special Operations Task Force Afghanistan. *Counterinsurgency Operations in Afghanistan, July to December 2004: Principles of Victory*. Combined Joint Special Operations Task Force Afghanistan, 2005.

Constable, Pamela. Gates Visits Kabul, Cites Rise in Cross-Border Attacks. *Washington Post*, January 17, 2007, p. A10.

Cordesman, Anthony H. *The Ongoing Lessons of Afghanistan: Warfighting, Intelligence, Force Transformation, and Nation Building*. Washington, D.C.: Center for Strategic and International Studies, 2004.

Corum, James. *Training Indigenous Forces in Counterinsurgency: A Tale of Two Insurgencies*. Carlisle, Pa.: U.S. Army War College, 2006.

Courtney, Morgan, Frederick Barton, and Bathsheba Crocker. *In the Balance: Measuring Progress in Afghanistan*. Washington, D.C.: Center for Strategic and International Studies, 2005.

Crumpton, Henry A. Intelligence and War: Afghanistan 2001–2002. In *Transforming U.S. Intelligence*, edited by Jennifer E. Sims and Burton Gerber. Washington, D.C.: Georgetown University Press, 2005.

Davis, Anthony. Kabul's Security Dilemma. *Jane's Defence Weekly*, Vol. 37, No. 24 (June 12, 2002): pp. 26–27.

————. Afghan Security Deteriorates as Taliban Regroup. *Jane's Intelligence Review*, Vol. 15, No. 5 (May 2003): pp. 10-15.

Dayan, General Moshe. Folder 488, 1-31 May 1968, #32 History File. Box 34, Record Group 319, National Archives and Records Administration.

Deady, Timothy K. Lessons from a Successful Counterinsurgency: The Philippines, 1899–1902. *Parameters*, Vol. XXXV, No. 1 (Spring 2005).

Denoeux, Guilain. The Forgotten Swamp: Navigating Political Islam. *Middle East Policy*, Vol. 9, No. 2 (June 2002).

Dobbins, James, Seth G. Jones, Keith Crane, and Beth Cole DeGrasse. *The Beginner's Guide to Nation-Building.* Santa Monica, Calif.: RAND Corporation, MG-557-SRF, 2007. As of January 19, 2007:
http://www.rand.org/pubs/monographs/MG557/

Dobbins, James, Seth G. Jones, Keith Crane, Andrew Rathmell, Brett Steele, Richard Teltschik, and Anga Timilsina. *America's Role in Nation Building: From Germany to Iraq.* Santa Monica, Calif: RAND Corporation, MR-1753-RC, 2003. As of January 19, 2007:
http://www.rand.org/pubs/monograph_reports/MR1753

Dunbar, Charles. Afghanistan in 1986: The Balance Endures. *Asian Survey*, Vol. 27, No. 2: pp. 127–142

Edelstein, David M. Occupational Hazards: Why Military Occupations Succeed or Fail. *International Security*, Vol. 29, No. 1 (Summer 2004).

Eikenberry, LTG Karl. *Statement of Lt. Gen. Karl Eikenberry, Commander, Combined Forces Command—Afghanistan, Testimony Before the House Armed Services Committee.* June 28, 2006.

Einhorn, Jessica. "The World Bank's Mission Creep," *Foreign Affairs*, Vol. 80, No. 5 (2001): pp. 22–35.

Etzioni, Amitai. *From Empire to Community: A New Approach to International Relations.* New York: Palgrave Macmillan, 2004.

————. A Self-Restrained Approach to Nation-Building by Foreign Powers. *International Affairs*, Vol. 80, No. 1 (2004).

Fearon, James. *Testimony to U.S. House of Representatives, Committee on Government Reform, Subcommittee on National Security, Emerging Threats, and International Relations on "Iraq: Democracy or Civil War?"* September 13, 2006.

Fearon, James D., and David D. Laitin. Explaining Interethnic Cooperation. *American Political Science Review*, Vol. 90, No. 4 (December 1996): pp. 715–735.

————. Ethnicity, Insurgency, and Civil War. *American Political Science Review*, Vol. 97, No. 1 (February 2003): pp. 75–90.

Fighting in Afghanistan Leaves 40 Insurgents Dead. American Forces Press Service, June 22, 2005.

Franks, GEN Tommy. *American Soldier*. New York: HarperCollins, 2004.

Fukuyama, Francis. *State-Building: Governance and World Order in the 21st Century*. Ithaca, N.Y.: Cornell University Press, 2004.

Gall, Carlotta and David Rohde. Militants Escape Control of Pakistan, Officials Say. *New York Times*, January 15, 2007, p. A1.

Gall, Carlotta, and Ismail Khan. Taliban and Allies Tighten Grip in Northern Pakistan. *New York Times*, December 11, 2006: pp. A1, A17.

Galula, David. *Counterinsurgency Warfare: Theory and Practice*. St. Petersburg, Fla.: Hailer Publishing, 2005.

George, Alexander L. Case Studies and Theory Development: The Method of Structured, Focused Comparison. In Paul Gordon Lauren, ed.*, Diplomacy: New Approaches in History, Theory, and Policy*. New York: Free Press, 1979.

George, Alexander L., and Timothy J. McKeown. Case Studies and Theories of Organizational Decision Making. In *Advances in Information Processing in Organizations: A Research Annual*, Vol. 2, edited by Robert F. Coulam and Richard A. Smith. Greenwich, Conn.: JAI Press, 1985.

Goldthorpe, John H. Current Issues in Comparative Macrosociology: A Debate on Methodological Issues. In Mjoset and Engelstad, 1997.

Government of Afghanistan. *Security Sector Reform: Disbandment of Illegal Armed Groups Programme (DIAG) and Government of Afghanistan, Disarmament, Demobilisation, and Reintegration Programme (DDR)*. Kabul: Government of Afghanistan, October 2005.

Government of Germany. *Doha II Conference on Border Management in Afghanistan: A Regional Approach*. Berlin: Government of Germany, 2006.

Government of Germany, Federal Foreign Office and Federal Ministry of Interior. *Assistance in Rebuilding the Police Force in Afghanistan*. Berlin, Germany: Federal Foreign Office and Federal Ministry of the Interior, March 2004.

Grare, Frederic. *Pakistan: The Resurgence of Baluch Nationalism*. Washington, D.C.: Carnegie Endowment for International Peace, January 2006.

Grau, Lester, ed. *The Bear Went Over the Mountain: Soviet Combat Tactics in Afghanistan*. Washington, D.C.: National Defense University Press, 1996.

—————. *Artillery and Counterinsurgency: The Soviet Experience in Afghanistan*. Fort Leavenworth, Kan.: Foreign Military Studies Office, 1997.

Gunmen Set Fire to Schools in Ghazni, Kandahar Provinces. Pajhwok Afghan News, December 24, 2005.

Haass, Richard N. *Intervention: The Use of American Military Force in the Post-Cold War World*. Washington, D.C.: Carnegie Endowment for International Peace, 2004.

Hastert, Paul L. Operation Anaconda: Perception Meets Reality in the Hills of Afghanistan. *Studies in Conflict and Terrorism*, Vol. 28, No.1 (January–February 2005).

Herd, Walter. *World War III: The Global Unconventional War on Terror*. Fort Bragg, N.C.: United States Army Special Operations Command, 2005.

Herd, COL Walter M., COL Patrick M. Higgins, LT COL Adrian T. Bogart III, MAJ A. Davey, and CAPT Daudshah S. Andish. *One Valley at a Time*. Fort Bragg, N.C.: Combined Joint Special Operations Task Force–Afghanistan, 2005.

Hersh, Seymour M. The Other War: Why Bush's Afghanistan Problem Won't Go Away. *The New Yorker* (April 12, 2004).

Hill, Luke. NATO to Quit Bosnia, Debates U.S. Proposals. *Jane's Defence Weekly*, Vol. 40, No. 23 (December 10, 2003): p. 6.

Hironaka, Ann. *Neverending Wars: The International Community, Weak States, and the Perpetuation of Civil War*. Cambridge, Mass.: Harvard University Press, 2005.

Hoffman, Bruce. *Insurgency and Counterinsurgency in Iraq*. Santa Monica, Calif.: RAND Corporation, OP-127-IPC/CMEPP, 2004. As of January 19, 2007: http://www.rand.org/pubs/occasional_papers/OP127/

———. *Inside Terrorism*, New York: Columbia University Press, 2006.

Hosmer, Stephen T. *The Army's Role in Counterinsurgency and Insurgency*. Santa Monica, Calif.: RAND Corporation, R-3947-A, 1990. As of January 19, 2007: http://www.rand.org/pubs/reports/R3947/

Huggins, Martha K. *Political Policing: The United States and Latin America*. Durham, N.C.: Duke University Press, 1998.

Huntington, Samuel P. *The Clash of Civilizations and the Remaking of World Order*. New York: Simon and Schuster, 1996.

International Crisis Group. *Afghanistan: Getting Disarmament Back on Track*. Kabul: International Crisis Group, 2005.

———. *Countering Afghanistan's Insurgency: No Quick Fixes*. Kabul: International Crisis Group, 2006.

International Institute for Strategic Studies. *The Military Balance, 1995/96*. London: Oxford University Press, 1995.

———. *The Military Balance, 1998/99*. London: Oxford University Press, 1998.

The International Republican Institute. *Afghanistan: Election Day Survey*. Kabul: The International Republican Institute, October 9, 2004.

International Security Assistance Force. *Opposing Militant Forces: Elections Scenario*. Kabul: ISAF, 2005.

Interview with Mullah Dadullah. Al Jazeera TV, July 2005.

Jackson, Brian A., John C. Baker, Peter Chalk, Kim Cragin, John V. Parachini, and Horacio Trujillo. *Aptitude for Destruction,* Vol. 1*: Organizational Learning in Terrorist Groups and Its Implications for Combating Terrorism*. Santa Monica, Calif.: RAND Corporation, MG-331-NIJ, 2005. As of January 19, 2007: http://www.rand.org/pubs/monographs/MG331/

————. *Aptitude for Destruction,* Vol. 2: *Case Studies of Organizational Learning in Five Terrorist Groups*. Santa Monica, Calif: RAND Corporation, MG-332-NIJ, 2005. As of January 19, 2007: http://www.rand.org/pubs/monographs/MG332/

Jalali, Ali A. Afghanistan: The Anatomy of an Ongoing Conflict. *Parameters*, Vol. XXXI, No. 1 (Spring 2001): p. 86.

————. The Future of Afghanistan. *Parameters*, Vol. XXXVI, No. 1 (Spring 2006).

Johnson, Ed. Gates Wants NATO to Reorganize Afghanistan Mission. Bloomberg News, December 12, 2007.

Jones, Seth G. Averting Failure in Afghanistan. *Survival*, Vol. 48, No. 1 (Spring 2006): pp. 111–128.

————. Pakistan's Dangerous Game. *Survival*, Vol. 49, No. 1, (Spring 2007): pp. 15–32.

Jones, Seth G., Jeremy M. Wilson, Andrew Rathmell, and K. Jack Riley. *Establishing Law and Order After Conflict*. Santa Monica, Calif.: RAND Corporation, MG-374-RC, 2005. As of January 19, 2007: http://www.rand.org/pubs/monographs/MG374/

Judah, Tim. The Taliban Papers. *Survival*, Vol. 44, No. 1 (Spring 2002): pp. 69–80.

Karzai Condemns Murder of Clerics. Pajhwok Afghan News, October 18, 2005.

Karzai, Hekmat. *Afghanistan and the Globalisation of Terrorist Tactics*. Singapore: Institute of Defence and Strategic Studies, January 2006.

————. *Afghanistan and the Logic of Suicide Terrorism*. Singapore: Institute of Defence and Strategic Studies, March 2006.

Karzai, Hekmat, and Seth G. Jones. How to Curb Rising Suicide Terrorism in Afghanistan. *Christian Science Monitor*, July 18, 2006.

Kaufmann, Daniel. Myths and Realities of Governance and Corruption. In *Global Competitiveness Report 2005–2006*. Geneva: World Economic Forum, 2005.

Kaufmann, Daniel, Aart Kraay, and Massimo Mastruzzi. *Governance Matters III: Governance Indicators for 1996–2002*. Washington, D.C.: World Bank, 2002.

———. *Governance Matters V: Aggregate and Individual Governance Indicators for 1996–2005*. Washington, D.C.: World Bank, 2006.

Kelly, Terrence K. *Options for Transitional Security Capabilities for America*. Santa Monica, Calif.: RAND Corporation, TR-353-A, 2005. As of January 19, 2007: http://www.rand.org/pubs/technical_reports/TR353/

Kepel, Gilles. *Jihad: The Trail of Political Islam*. Cambridge, Mass.: Harvard University Press, 2002.

Khalilzad, Zalmay. How to Nation-Build: Ten Lessons from Afghanistan. *National Interest*, No. 80 (Summer 2005): pp. 19–27.

———. Outgoing U.S. Envoy Enthusiastic About Afghanistan's Future. Interview on Sherberghan Jowzjan Aina Television, June 18, 2005.

Khattak, Iqbal. 40 Militants Killed in North Waziristan. *Daily Times* (Pakistan), September 30, 2005.

King, Gary, Robert Keohane, and Sidney Verba. *Designing Social Inquiry: Scientific Inference in Qualitative Research*. Princeton, N.J.: Princeton University Press, 1994.

Kitson, Frank. *Low Intensity Operations: Subversion, Insurgency and Peacekeeping*. London: Faber and Faber, 1971.

Koehler, Jan. *Conflict Processing and the Opium Poppy Economy in Afghanistan*. Jalalabad: Project for Alternative Livelihoods, June 2005.

Krepinevich, Andrew F., Jr. *The Army and Vietnam*. Baltimore, Md.: Johns Hopkins University Press, 1986.

———. How to Win in Iraq. *Foreign Affairs*, Vol. 84, No. 5 (September/October 2005).

MacDonald, Scott. Minister's Visit Hints at Taliban Split. Reuters, October 20, 2001.

Maley, William, ed. *Fundamentalism Reborn? Afghanistan and the Taliban*. New York: New York University Press, 2001.

Mani, Rama. *Ending Impunity and Building Justice in Afghanistan*. Kabul: Afghanistan Research and Evaluation Unit, 2003.

Manuel, Anja, and P. W. Singer. A New Model Afghan Army. *Foreign Affairs*, Vol. 81, No. 4 (July/August 2002): pp. 44–59.

Maples, LTG Michael D. *Current and Projected National Security Threats to the United States, Statement for the Record, Senate Armed Services Committee*. February 28, 2006.

McCaffrey, GEN Barry R., (ret.). Trip to Afghanistan and Pakistan. Memorandum from General McCaffrey to COL Mike Meese and COL Cindy Jebb, United States Military Academy, June 2006.

McClintock, Michael. *The American Connection*. London: Zed Books, 1985.

McConnell, J. Michael. Director of National Intelligence. *Annual Threat Assessment of the Director of National Intelligence for the Senate Select Committee on Intelligence*. Statement for the record, February 5, 2008.

McInerney, Michael. The Battle for Deh Chopan, Part 1. *Soldier of Fortune*, August 2004.

———. The Battle for Deh Chopan, Part 2. *Soldier of Fortune*, September 2004.

McNerney, Michael J., Stabilization and Reconstruction in Afghanistan: Are PRTs a Model or a Muddle? *Parameters*, Vol. XXXV, No. 4 (Winter 2005–2006).

Miller, Laurel, and Robert Perito. *Establishing the Rule of Law in Afghanistan*. Special Report 117. Washington, D.C.: United States Institute of Peace, 2004.

Mosjet, Lars, and Frederik Engelstad, eds. *Coparative Social Research*, Vol. 16: *Methodological Issues in Comparative Social Science*. Greenwich, Conn.: JAI Press, 1997.

Motlagh, Jason. Weapons Convoy from Iran Reported. *Washington Times*, June 22, 2007, p. A13.

Mulford, David C. *Afghanistan Has Made a Remarkable Transition*. New Delhi: U.S. Department of State, February 2006.

Nagl, John A. *Learning to Eat Soup with a Knife: Counterinsurgency Lessons from Malaya and Vietnam*. Chicago: University of Chicago Press, 2005.

National Commission on Terrorist Attacks upon the United States. *The 9/11 Commission Report: Final Report of the National Commission on Terrorist Attacks upon the United States*. New York: W. W. Norton, 2004.

National Intelligence Council. *The Terrorist Threat to the U.S. Homeland*. Washington, D.C.: National Intelligence Council, 2007.

Naylor, Sean. *Not a Good Day to Die: The Untold Story of Operation Anaconda*. New York: Berkley Books, 2005.

Negroponte, John D. *Annual Threat Assessment of the Director of National Intelligence for the Senate Armed Services Committee*. Statement for the record, February 28, 2006.

Oakley, Robert B., Michael J. Dziedzic, and Eliot M. Goldberg, eds. *Policing the New World Disorder: Peace Operations and Public Security*. Washington, D.C.: National Defense University Press, 1998.

Offices of Inspector General of the Departments of State and Defense. *Interagency Assessment of Afghanistan Police Training and Readiness.* Washington, D.C.: Offices of Inspector General of the Departments of State and Defense, 2006.

Orr, Robert C., ed. *Winning the Peace: An American Strategy for Post-Conflict Reconstruction.* Washington, D.C.: The CSIS Press, 2004.

Pagent, Julian. *Counter-Insurgency Campaigning.* London: Faber and Faber, 1967.

Pajhwok News Describes Video of Afghan Beheading by "Masked Arabs," Taliban. Pajhwok Afghan News, October 9, 2005.

Pakistani Law Enforcers Intensify Hunt for Haqqani. Pajhwok Afghan News, March 7, 2006.

Pakistan Strikes Suspected al Qaeda Camp. Associated Press, March 1, 2006.

Pape, Robert. *Dying to Win: The Strategic Logic of Suicide Terrorism.* New York: Random House, 2005.

Paris, Roland. *At War's End: Building Peace After Civil Conflict.* New York: Cambridge University Press, 2004.

Peace Pact: North Waziristan, n.p., September 5, 2006.

Perito, Robert M. *The American Experience with Police in Peace Operations.* Clementsport, Canada: The Canadian Peacekeeping Press, 2002.

———. *Where Is the Lone Ranger When We Need Him? America's Search for a Postconflict Stability Force.* Washington, D.C.: United States Institute of Peace, 2004.

Pirnie, Bruce R., Alan J. Vick, Adam Grissom, Karl P. Mueller, and David T. Orletsky. *Beyond Close Air Support: Forging a New Air-Ground Partnership.* Santa Monica, Calif.: RAND Corporation, MG-301-AF, 2005. As of January 19, 2007: http://www.rand.org/pubs/monographs/MG301/

"Pro-Karzai" Cleric Killed by Bomb in Mosque in Khost Province. Pajhwok Afghan News, October 14, 2005.

Ragin, Charles C. Comparative Sociology and the Comparative Method. *International Journal of Comparative Sociology*, Vol. 22, Nos. 1–2, (March–June 1981): pp. 102–120.

RAND-MIPT Incident Database.

Rashid, Ahmed. *Taliban: Militant Islam, Oil and Fundamentalism in Central Asia.* New Haven, Conn.: Yale University Press, 2000.

———. Pakistan and the Taliban. In Maley, 2001.

———. Who's Winning the War on Terror? *YaleGlobal* (September 5, 2003).

Redding, MAJ Robert W. 19th SF Group Utilizes MCA Missions to Train Afghan National Army Battalions. *Special Warfare*, Vol. 17 (February 2005): pp. 22–27.

Religious Scholars Call on Taliban to Abandon Violence. Pajhwok Afghan News, July 28, 2005.

Reuter, Christoph. *My Life Is a Weapon: A Modern History of Suicide Bombing.* Princeton, N.J.: Princeton University Press, 2004.

Roe, MAJ Andrew M. To Create a Stable Afghanistan. *Military Review* (November–December 2005).

Rohde, David. Foreign Fighters of Harsher Bent Bolster Taliban. *New York Times*, October 30, 2007, p. A1.

Rothstein, Hy S. *Afghanistan and the Troubled Future of Unconventional Warfare.* Annapolis, Md.: Naval Institute Press, 2006.

Roy, Olivier. *Islam and Resistance in Afghanistan*, 2nd ed. New York: Cambridge University Press, 1990.

Rubin, Barnett R. *The Search for Peace in Afghanistan: From Buffer State to Failed State.* New Haven, Conn.: Yale University Press, 1995.

―――. *The Fragmentation of Afghanistan: State Formation and Collapse in the International System.* New Haven, Conn.: Yale University Press, 2002.

―――. (Re)Building Afghanistan. *Current History*, Vol. 103, No. 672 (April 2004): pp. 165–170.

―――. *Road to Ruin: Afghanistan's Booming Opium Industry.* New York: Center on International Cooperation, October 2004.

―――. *Afghanistan and the International Community: Implementing the Afghanistan Compact.* New York: Council on Foreign Relations, 2006.

―――. *Afghanistan's Uncertain Transition from Turmoil to Normalcy.* New York: Council on Foreign Relations, 2006.

―――. Still Ours to Lose: Afghanistan on the Brink. Submitted as written testimony to the Senate Foreign Relations Committee, September 21, 2006.

―――. Saving Afghanistan. *Foreign Affairs*, Vol. 86, No. 1, January/February 2007.

Rubin, Barnett R., and Andrea Armstrong. Regional Issues in the Reconstruction of Afghanistan. *World Policy Journal*, Vol. XX, No. 1 (Spring 2003).

Sageman, Marc. *Understanding Terror Networks.* Philadelphia: University of Pennsylvania Press, 2004.

Saleh, Amrullah. *Strategy of Insurgents and Terrorists in Afghanistan.* Kabul: National Directorate for Security, 2006.

Sargent, MAJ Ron. Strategic Scouts for Strategic Corporals. *Military Review*, Vol. 85, No. 2 (March–April 2005).

Schroen, Gary. *First In: An Insider's Account of How the CIA Spearheaded the War on Terror in Afghanistan*. New York: Ballantine Books, 2005.

Schweller, Randall L. Managing the Rise of Great Powers: History and Theory. In Alastair Iain Johnston and Robert S. Ross, eds., *Engaging China: The Management of an Emerging Power*. New York: Routledge, 1999.

Sedra, Mark. *Challenging the Warlord Culture: Security Sector Reform in Post-Taliban Afghanistan*. Bonn, Germany: Bonn International Center for Conversion, 2002.

The Senlis Group. *Helmand at War: The Changing Nature of the Insurgency in Southern Afghanistan and Its Effects on the Future of the Country*. London: The Senlis Council, 2006.

Shahzad, Syed Saleem. Taliban's New Commander Ready for a Fight. *Asia Times*, May 20, 2006.

Shaikh, Najmuddin A. Worsening Ties with Kabul. *Dawn* (Pakistan), December 13, 2006.

Simpson, Charles. *Inside the Green Berets: The First Thirty Years*. Novato, Calif.: Presidio Press, 1982.

Singer, P. N. *Corporate Warriors: The Rise of the Privatized Military Industry*. Ithaca, N.Y.: Cornell University Press, 2003.

Skocpol, Theda, and Margaret Somers. The Uses of Comparative History in Macrosocial Inquiry. *Comparative Studies in Society and History*, Vol. 22, No. 2 (1980): pp. 174–197.

Spokesman Rejects Afghan Government's Amnesty Offer for Taliban Leader. Peshawar Afghan Islamic Press, May 9, 2005.

Spokesman Says Taliban "Fully Organized." *Islamabad Ausaf*, June 23, 2005: pp. 1, 6.

Stanley, William. Building New Police Forces in El Salvador and Guatemala: Learning and Counter-Learning. In Tor Tanke Holm and Espen Barth Eide, eds., *Peacebuilding and Police Reform*. Portland, Oreg.: Frank Cass, 2000.

Starr, S. Frederick. *U.S. Afghanistan Policy: It's Working*. Washington, D.C.: Central Asia-Caucasus Institute, Johns Hopkins University, 2004.

————. Sovereignty and Legitimacy in Afghan Nation-Building. In Francis Fukuyama, ed., *Nation-Building: Beyond Afghanistan and Iraq*. Baltimore, Md.: Johns Hopkins University Press, 2006.

Stockholm International Peace Research Institute. *SIPRI Yearbook 1991: World Armaments and Disarmament*. New York: Oxford University Press, 1991.

Straziuso, Jason. U.S. Casualties in Afghanistan Hit Record. *Navy Times*, January 2, 2008.

Swami, Praveen. Covert Contestation. *Frontline*, Vol. 22, No. 19 (September 2005).

Taliban Claim Attack on Police in Jalalabad, Nangarhar Province. Kabul National TV, January 7, 2006.

Taliban Claim Killing of Pro-Government Religious Scholars in Helmand. Afghan Islamic Press, July 13, 2005.

Taliban Claim Responsibility for Killing Afghan Cleric. Kabul Tolu Television, May 29, 2005.

Taliban Claim Responsibility for Suicide Bomb Attack in Afghan Kandahar Province. Peshawar Afghan Islamic Press, October 9, 2005.

Taliban Claim Shooting Down U.S. Helicopter. *The News* (Islamabad), June 29, 2005.

Taliban Execute Afghan Woman on Charges of Spying for U.S. Military. Afghan Islamic Press, August 10, 2005.

Taliban Launch Pirate Radio Station in Afghanistan. Agence France Presse, April 18, 2005.

Taliban Military Chief Threatens to Kill U.S. Captives, Views Recent Attacks, Al-Qa'ida. Interview with Al Jazeera TV, July 18, 2005.

Taliban Says Responsible for Pro-Karzai Cleric's Killing, Warns Others. *The News* (Islamabad), May 30, 2005.

Taliban Spokesman Condemns Afghan Parliament as "Illegitimate." Sherberghan Aina TV, December 19, 2005.

Taliban Threatens Teachers, Students in Southern Afghan Province. Pajhwok Afghan News, January 3, 2006.

Tandy, Karen P. *Statement of Karen P. Tandy, Administrator, U.S. Drug Enforcement Agency, Testimony Before the House Armed Services Committee*. June 28, 2006.

Tapper, Richard. Anthropologists, Historians, and Tribespeople on Tribe and State Formation in the Middle East. In Philip S. Khoury and Joseph Kostiner, eds., *Tribes and State Formation in the Middle East*. Berkeley, Calif.: University of California Press, 1990.

Tarzi, Amin. Afghanistan: Kabul's India Ties Worry Pakistan. Radio Free Europe/ Radio Liberty, April 16, 2006.

Tilly, Charles. *The Formation of National States in Western Europe*. Princeton, N.J.: Princeton University Press, 1975.

———. Means and Ends of Comparison in Macrosociology. In Mjoset and Engelstad, 1997.

Tomes, Robert R. Relearning Counterinsurgency Warfare. *Parameters*, Vol. XXXIV, No. 1 (Spring 2004).

Transparency International. *Global Corruption Report 2006*. Berlin: Transparency International, 2006.

Trinquier, Roger. *Modern Warfare: A French View of Counterinsurgency*. Trans. Daniel Lee. New York: Praeger, 1964.

Tse-Tung, Mao. *Selected Military Writings of Mao Tse-Tung*. Peking, China: Foreign Language Press, 1963.

UK Source in Afghanistan Says al Qaeda Attacks Boost Fear of Taliban Resurgence. *Guardian* (London), June 20, 2005.

United Nations. *United Nations Security Council Resolution 1383*. S/RES/1383, December 6, 2001.

———. *Report of the Secretary-General on the Situation in Afghanistan and Its Implications for International Peace and Security*. UN doc A/56/875-S/2002/278, March 18, 2002.

———. *Afghanistan: Opium Survey 2003*. Vienna: United Nations Office on Drugs and Crime, 2003.

———. *The Opium Economy in Afghanistan: An International Problem*. New York: United Nations Office on Drugs and Crime, 2003.

———. *Afghanistan Opium Survey 2004*. Vienna: United Nations Office on Drugs and Crime, 2004.

———. *World Drug Report*. New York: United Nations Office on Drugs and Crime, 2004.

———. *Afghanistan: Opium Survey 2005*. Kabul: United Nations Office on Drugs and Crime, 2005.

———. *Afghanistan Opium Survey 2006*. Kabul: United Nations Office on Drugs and Crime, 2006.

———. *Afghanistan Opium Survey 2007*. Kabul: United Nations Office on Drugs and Crime, 2007.

United Nations Assistance Mission in Afghanistan. Agreement on Provisional Arrangements in Afghanistan Pending the Reestablishment of Permanent Government Institutions, December 2001.

———. *Suicide Attacks in Afghanistan: 2001–2007*. Kabul: United Nations Assistance Mission to Afghanistan, 2007.

U.S. Air Force. Office of Air Force Lessons Learned (AF/XOL). *Operation Anaconda: An Air Power Perspective*. Washington, D.C.: Headquarters United States Air Force AF/XOL, February 2005. As of January 19, 2007: http://www.af.mil/shared/media/document/AFD-060726-037.pdf

U.S. Army Training and Doctrine Command. *Operation Enduring Freedom: Tactics, Techniques, and Procedures.* Fort Leavenworth, Kan.: U.S. Army Training and Doctrine Command, December 2003.

————. *Observations and Lessons Learned: Task Force Devil, 1st Brigade Combat Team, 82 Airborne Division.* Fort Leavenworth, Kan.: U.S. Army Training and Doctrine Command, January 2004.

U.S. Department of Defense. *Department of Defense Dictionary of Military and Associated Terms.* Joint Publication 1-02. Washington, D.C.: U.S. Department of Defense, 2001.

————. *Information Operations Roadmap.* Washington, D.C.: U.S. Department of Defense, 2003.

U.S. Department of State. *Capitol Hill Monthly Update, Afghanistan.* Washington, D.C.: U.S. Department of State, June 2004.

————. *Border Management Initiative: Information Brief.* Kabul: Afghanistan Reconstruction Group, U.S. Department of State, 2005.

U.S. Government Accountability Office. *Afghanistan Security: Efforts to Establish Army and Police Have Made Progress, but Future Plans Need to Be Better Defined.* Washington, D.C.: GAO, 2005.

U.S. House of Representatives, Committee on International Relations and U.S. Senate, Committee on Foreign Relations. *Legislation on Foreign Relations Through 2000.* Washington, D.C.: U.S. Government Printing Office, 2001.

U.S. Marine Corps. *Small Wars Manual.* Washington, D.C.: U.S. Government Printing Office, 1940.

————. *After Action Report on Operations in Afghanistan.* Camp Lejeune, N.C.: United States Marine Corps, August 2004.

U.S. Senate. *Afghanistan Stabilization and Reconstruction: A Status Report—Hearing Before the Committee on Foreign Relations.* S.Hrg. 108-460, January 27, 2004.

U.S. Marine Corps. *After Action Report on Operations in Afghanistan.* Camp Lejeune, N.C.: United States Marine Corps, August 2004.

Van Evera, Stephen. *Guide to Methods for Students of Political Science.* Ithaca, N.Y.: Cornell University Press, 1997.

Walsh, Declan. Pakistan: Resurgent al-Qaida Plotting Attacks on West From Tribal Sanctuary. *Guardian* (London), September 27, 2007, p. 24.

Walt, Stephen M. *Taming American Power: The Global Response to U.S. Primacy.* New York: W. W. Norton, 2005.

Weber, Max. Politics as a Vocation. In H. H. Gerth and C. Wright Mills, eds., *From Max Weber: Essays in Sociology.* New York: Oxford University Press, 1958.

The White House. *Rebuilding Afghanistan*. Washington, D.C.: The White House, 2004.

Wilensky, Robert J. *Military Medicine to Win Hearts and Minds: Aid to Civilians in the Vietnam War*. Lubbock, Tex.: Texas Tech University Press, 2004.

Woodward, Bob. *Bush at War*. New York: Simon and Schuster, 2002.

World Bank. *Reforming Public Institutions and Strengthening Governance*. Washington D.C.: World Bank, 2000.

———. *Afghanistan: Managing Public Finances for Development*. Washington, D.C.: World Bank, 2005.

———. *Aggregate Governance Indicators Dataset, 1996–2006*. Washington, D.C.: World Bank, 2007.

Wright, Robin. Iranian Destined for Taliban Seized in Afghanistan. *Washington Post*, September 16, 2007, p. A19.

Yousafzai, Sami, and Ron Moreau. Unholy Allies. *Newsweek*, September 26, 2005: pp. 40–42.

Zisk, Kimberly Marten. *Enforcing the Peace: Learning from the Imperial Past*. New York: Columbia University Press, 2004.

Zogby, James. *2005 Arab Attitudes Toward U.S.: Good News and Bad News*. New York: Zogby International, November 7, 2005.

Zulfiqar, Shahzada. Endless War. *The Herald* (Pakistan), April 2006: pp. 33–36.

About the Author

Seth G. Jones is a Political Scientist at the RAND Corporation and an Adjunct Professor at Georgetown University's Edmund A. Walsh School of Foreign Service. He is the author of *In the Graveyard of Empires: America's War in Afghanistan* (W. W. Norton, forthcoming) and *The Rise of European Security Cooperation* (Cambridge University Press, 2007). He has published articles on a range of national security subjects in *International Security, The National Interest, Political Science Quarterly, Security Studies, Chicago Journal of International Law, International Affairs,* and *Survival,* as well as such newspapers and magazines as the *New York Times, Newsweek,* the *Financial Times,* and the *International Herald Tribune.* He received his MA and PhD from the University of Chicago.